Limper
Mixing of Rubber Compounds

Andreas Limper

Mixing of Rubber Compounds

Hanser Publishers, Munich

HANSER

Hanser Publications, Cincinnati

The Editor:
Prof. Dr.-Ing. Andreas Limper
Harburg-Freudenberger Maschinenbau GmbH, Asdorfer Strasse 60, 57258 Freudenberg, Germany

Distributed in North and South America by:
Hanser Publications
6915 Valley Avenue, Cincinnati, Ohio 45244-3029, USA
Fax: (513) 527-8801
Phone: (513) 527-8977
www.hanserpublications.com

Distributed in all other countries by
Carl Hanser Verlag
Postfach 86 04 20, 81631 München, Germany
Fax: +49 (89) 98 48 09
www.hanser.de

The use of general descriptive names, trademarks, etc., in this publication, even if the former are not especially identified, is not to be taken as a sign that such names, as understood by the Trade Marks and Merchandise Marks Act, may accordingly be used freely by anyone.
While the advice and information in this book are believed to be true and accurate at the date of going to press, neither the authors nor the editors nor the publisher can accept any legal responsibility for any errors or omissions that may be made. The publisher makes no warranty, express or implied, with respect to the material contained herein.

Library of Congress Cataloging-in-Publication Data

Limper, A. (Andreas)
  Mixing of rubber compounds / Andreas Limper.
    p. cm.
  Includes bibliographical references and index.
  ISBN 978-1-56990-458-9 (hardcover) -- ISBN 1-56990-458-8 (hardcover) -- ISBN 978-3-446-41743-4 (hardcover) 1. Rubber. 2. Rubber machinery. 3. Mixing machinery. I. Title.
  TS1890.L55 2012
  678'.20284--dc23
                      2011047325

Bibliografische Information Der Deutschen Bibliothek
Die Deutsche Bibliothek verzeichnet diese Publikation in der Deutschen Nationalbibliografie; detaillierte bibliografische Daten sind im Internet über <http://dnb.d-nb.de> abrufbar.

ISBN 978-3-446-41743-4

All rights reserved. No part of this book may be reproduced or transmitted in any form or by any means, electronic or mechanical, including photocopying or by any information storage and retrieval system, without permission in writing from the publisher.

© Carl Hanser Verlag, Munich 2012
Production Management: Steffen Jörg
Coverconcept: Marc Müller-Bremer, www.rebranding.de, München
Coverdesign: Stephan Rönigk
Typeset, printed, and bound by Kösel, Krugzell
Printed in Germany

Preface

Mixing of rubber compounds is a multi-discipline task. This book is based on a course on the compounding of elastomers. It describes the machine aspects, the relevant processes in a batch mixer, the polymer-specific requirements and the use of the internal mixer as a reactor. As the mixing process itself has a severe influence on the final product properties, the correlation of product quality and the outcome of the compounding process are widely discussed. Also the scientific background of the dispersion of fillers as well as the relevant conditions in physical and chemical terms are described in depth.

With this volume the authors offer on the one hand a well-rounded information package to beginners in the area of rubber mixing. On the other hand, experienced compounders should be able to deepen as well as widen their horizons of expertise. As the variety of rubber compounds is infinite, the authors concentrated on basic processes. Explanations of certain mechanisms, however, are also demonstrated including typical practical examples and their results.

As editor, I thank all authors for their investment of a great amount of private time into this work. I must also thank Mrs. Monika Stueve, Dr. Christine Strohm and Dr. Harald Sambale for their endurance, as the finalizing of this book took much longer than originally planned. Last but not least I thank my family and especially my wife for bearing with office work instead of leisure time.

Freudenberg, Germany *Andreas Limper*
January 2012

Preface

Table of Contents

1	**Internal Mixer – Configuration and Design**		1
	Dieter Berkemeier		
1.1	Machine Design		1
	1.1.1	Overall Features	1
	1.1.2	Mixing Chamber	2
1.2	Types of Internal Mixers		3
	1.2.1	Tangential Rotors	3
	1.2.2	Intermeshing Rotors	5
	1.2.3	Technological Comparison	5
	1.2.4	Intermeshing Rotors with Variable Clearance (VIC)	7
	1.2.5	Tandem Mixing	8
1.3	Feeding Hopper		11
	1.3.1	Design Considerations	11
	1.3.2	Pneumatic Feeding Hopper	12
	1.3.3	Hydraulic Feeding Hopper	13
	1.3.4	Comparative Aspects	14
1.4	Digital Ram Position Control		16
1.5	Mixing Chamber		18
	1.5.1	Hard-Coating	19
	1.5.2	Dust Sealing	21
	1.5.3	Spring Loaded Dust Seals	22
	1.5.4	Hydraulic Dust Seals with Yoke	23
	1.5.5	Hydraulic Dust Seals with Cylinders (CH)	25
	1.5.6	Comparison of Spring Loaded and Hydraulic Dust Seal Systems	25
1.6	Temperature Sensor		26
1.7	Plasticizer Oil Injection		29
1.8	Rotors		30
	1.8.1	Assembly and Cooling	30
	1.8.2	Rotor Bearings	31

	1.8.3	Rotors for Tangential Internal Mixers	32
		Basics	32
		Two-Wing Rotors	33
		Four-Wing rotors	33
		N-Rotor (Normal Rotor; also Called Standard-Rotor)	34
		Full-4-Wing (F4W)-Rotor	34
		ST®-Rotor (Synchronous Technology)	35
		ZZ 2-Rotor	36
		HDSC-Rotor (High Dispersion Super Cooled)	37
		MDSC-Rotor (Maximum Dispersion Super Cooled)	38
		Six-Wing Rotor	38
	1.8.4	Rotors for Intermeshing Internal Mixers	39
		History of Development	39
		Interlocking Technology	40
		PES-Technology	41
1.9	Mixer Base Plate		43
	1.9.1	Design	43
	1.9.2	Drop Door and Latch Assembly	44
	1.9.3	Drop Door and Toggle	45
References			46

2	**Processing Aspects of Rubber Mixing**	**47**
	A. Limper	
2.1	Mixing Principles	47
2.2	Process Description	50
2.3	Influence of Raw Material Properties	55
2.4	Influences of Process Parameters	57
2.5	Basic Considerations for the Development of a Mixing Cycle	66
References		68

3	**Mixing Characteristics of Polymers in an Internal Mixer**	**71**
	M. Rinker, A. Limper	
3.1	Natural Rubber (NR)	71
3.2	Ethylene Propylene Diene Rubber (EPDM)	75
3.3	Chloroprene Rubber (CR)	78
3.4	Styrene Butadiene Rubber (SBR)	80
3.5	Butadiene Rubber (BR)	84
3.6	(Acryl)Nitrile Butadiene Rubber NBR	84

3.7	Butyl Rubber	86
3.8	Fluor Rubber	86
3.9	Resins	87
3.10	General Considerations	88
References		93

4 Internal Mixer – a Reaction Vessel 95
Oliver Klockmann

4.1	The Silica Network	96
4.2	Influence of Mixing Time and Temperature on Hydrophobation	97
4.3	Chemistry of the Silica-Silane Reaction	99
4.4	Temperature Limits	101
4.5	Summary and Consequences	103
References		104

5 Effect of Process Parameters on Product Properties 107
Dr.-Ing. Harald Keuter

5.1	Introduction		107
	5.1.2	Quality Parameters of Raw Material	108
		5.1.2.1 Quality Parameters of EPDM Polymers	108
		5.1.2.2 Quality Parameters of Carbon Black	109
5.2	Raw Material Changes in the Rubber Mixing Room		116
	5.2.1	Increase of Carbon Black Fines Content during the Conveying Process	117
	5.2.2	Baking Behavior of Carbon Black in Conveying Pipes	118
5.3	Effect of Variations in Raw Material Quality Parameters on the Mixing Process		121
	5.3.1	EPDM Long Chain Branching	121
	5.3.2	Carbon Black Fines Content	124
	5.3.3	Carbon Black Pellet Hardness	125
5.4	Delivery Form of Sulphur		127
5.5	Weighing Accuracy		129
5.6	Predicting Product Quality		131
5.7	The Quality Assurance Concept "Future Mixing Room"		134
References			139

5.8	Rubber Compounding and its Impact on Product Properties *Dr.-Ing. Peter Ryzko*	141
5.9	Testing Methods for Rubber Compounds	144
	5.9.1 Mooney Viscometer	145
	5.9.2 Vulcameter	146
	5.9.3 Rubber Process Analyzer	147
	5.9.4 Carbon Black Dispersion Measurement	149
5.10	Factors Influencing Rubber Part Properties	149
5.11	The Mixing Process	151
	5.11.1 The Mixing Process and its Tasks	151
	5.11.2 Further Processes	153
5.12	Factors Influencing the Mixing Process	154
	5.12.1 Influence of Plasticizer Addition on the Mixing Process and Compound Properties	155
	5.12.2 Influences of the Mixing Process on the Injection Molding Process	157
	5.12.3 Influence of the Mixing Process on Extrusion	160
	5.12.4 Influence of the Milling Process on Compound and Part Properties	162
5.13	Summary	169
References		170

6	**Dispersion and Distribution of Fillers** *R. H. Schuster*	**173**
6.1	Dispersive and Distributive Mixing	174
	6.1.1 Dispersive Mixing	175
	6.1.2 Distributive Mixing	177
	6.1.3 Quality or "Goodness" of Mixes	178
6.2	Mechanism of Filler Dispersion	179
	6.2.1 Theoretical Approach	180
	6.2.2 Phases of Mixing Process	181
	6.2.3 Polymer-Filler versus Filler-Filler Interactions	183
6.3	Dispersion Measurements	189
	6.3.1 Macro-Dispersion	190
	6.3.1.1 Optical Transmission Microscopy	190
	6.3.1.2 Optical Roughness Measurements	191
	6.3.1.3 Mechanical Scanning Microscopy	192
	6.3.1.4 Reflectometry	193

	6.3.2	Micro-Dispersion	193
		6.3.2.1 Electrical Measurements	193
		6.3.2.2 Transmission Electron Microscopy	194
		6.3.2.3 Atomic Force Microscopy	195
6.4	Control of Dispersion by Process Parameters		196
	6.4.1	Mixing Procedures	196
	6.4.2	Temperature, Torque, and Power Consumption	197
	6.4.3	Mixing Time and Rotor Speed	198
	6.4.4	Cooling	199
	6.4.5	Alternative Dispersion Techniques	199
6.5	Materials Influences on Filler Dispersion		200
	6.5.1	Influence of the Polymer	201
		6.5.1.1 Adsorption from Solution and Melt	201
		6.5.1.2 Influence of the Polymer on Filler Dispersion	203
	6.5.2	Influence of the Filler Morphology and Surface Properties	205
		6.5.2.1 Influence of Surface Specific Area	205
		6.5.2.2 Influence of Structure	206
		6.5.2.3 Influence of Filler Surface Activity	207
		6.5.2.4 Dispersion Kinetics	209
		6.5.2.5 Filler Re-Agglomeration	211
	6.5.3	Influence of Oil on Filler Dispersion	211
6.6	Effects of Filler Dispersion on Material Behavior		212
	6.6.1	Effects on Rheological Properties	213
	6.6.2	Effects on Dynamical-Mechanical Properties	215
		6.6.2.1 Influence of Loading	216
		6.6.2.2 Strain Dependency	217
		6.6.2.3 Effect of Filler Surface Modification	218
	6.6.3	Effect on Ultimate Properties	219
6.7	Filler Distribution in Polymer Blends		222
	6.7.1	Compatibility of Rubbers	222
	6.7.2	Filler Partition	224
	6.7.3	Evaluation of Filler Distribution	224
	6.7.4	Distribution in Blends with Different Polymer Polarity	226
	6.7.5	Filler Distribution in Blends with Similar Polarity	227
	6.7.6	Filler Transfer	228
	6.7.7	Effects of Filler Distribution	229
References			230
Index			237

1 Internal Mixer – Configuration and Design

Dieter Berkemeier

■ 1.1 Machine Design

1.1.1 Overall Features

Irrespective of the way a mixing room is laid out, the internal mixer is the heart of the installation. The production capacity of a mixing line and the quality of the compounds produced are determined by the size of the internal mixer and its mixing efficiency. The machine is modular in construction and consists of three major sub-assemblies (Fig. 1.1).

The upper part of the mixer is the feeding hopper. Raw ingredients are fed into the mixer via this unit. Inside the feeding hopper is a ram which presses the raw

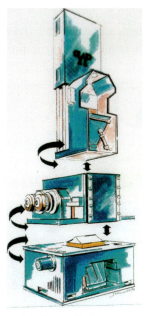

Figure 1.1 Modular design of a modern internal mixer

ingredients into the mixing chamber located beneath it. The actual mixing process takes place in the central part of the internal mixer, the mixing chamber. The base plate is situated below the mixing chamber and contains a discharge door in order to empty the mixer after the completion of the mixing process.

The three major sub-assemblies, the feeding hopper, mixing chamber, and base plate can each be turned separately through 180° and can be assembled to create the most suitable mixer configuration for any particular installation conditions, the position of material feeding systems, or downstream equipment. If necessary, this configuration can be changed at any time for mixer relocation or mill room reorganization.

1.1.2 Mixing Chamber

Similar to the complete mixer, the modular design concept is also applied to the mixing chamber (Fig. 1.2). It consists of two mixer end frames, two mixing chamber halves, and a set of two rotors. The mixer end frames with the rotor end plates form the axial limits of the mixing chamber and at the same time provide the housings for the rotor bearings. They are of split design and thus make the dismantling of the mixer easier, for example, when replacing the rotor.

The mixing chamber halves are installed between the two end frames. These define the radial limits of the mixing chamber and consist of two semi cylindrical shells, which are strengthened by additional supports welded onto their external surfaces.

The two rotors are located inside the mixing chamber. These have basically a cylindrical construction with several wings or protrusions attached diagonally

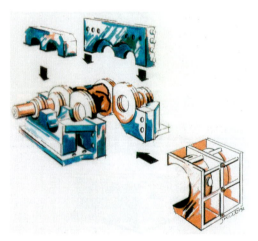

Figure 1.2 Modular design of the mixing chamber

onto the cylindrical rotor body. The rotors move the compound around within the mixing chamber. They rotate in opposite directions and thus draw the compound in, under the feeding ram, or carry it back to the ram from the area of the discharge door.

In this way the compound is moved in both axial and radial directions inside the mixing chamber.

At the top, the mixing chamber is closed by the ram (housed in the feeding hopper) and at the bottom by the discharge door (housed in the base plate). The discharge door (also known as the drop door) is firmly locked shut during the whole mixing cycle. In contrast, the pneumatic or hydraulic powered ram is not fixed and is pushed down onto the compound with a constant pressure. In this way the ram has the effect of preventing large load peaks during the mixing process.

New developments also allow a "ram position control" in certain mixing phases, in which the ram pressure is varied. However, this is possible only for hydraulic rams and requires sophisticated process control (see Chapter 2).

All components in contact with the compound, i.e., the ram, mixing chamber halves, rotor end plates, rotors, and discharge door, can be temperature controlled and cooled. For this purpose drilled holes or cast chambers are incorporated into these components. Cooling is necessary to enable the efficient removal of heat generated in the compound during the mixing process and to prevent hot spots, which may lead to partially sticking/scorching material portions.

■ 1.2 Types of Internal Mixers

There are two basic types of internal mixers characterized by the rotor system they use:

- Tangential rotors
- Intermeshing rotors

1.2.1 Tangential Rotors

The distinguishing characteristic of this type of machine (Fig. 1.3) is the fact that the movement patterns of the wing tips of each rotor do not touch each other. This means that, as the rotors rotate, a gap exists between the wing tips of the two rotors. The speed of each rotor can therefore be independently controlled and the rotors can run at different speeds. In practice, one rotor normally turns at a speed approximately 10% higher than the other. A relatively recent variation is to run

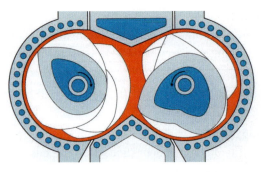

Figure 1.3 Sectional drawing of an internal mixer with tangential rotors

both rotors at the same speed. The English term for this is "even speed mixing" and with the correct radial adjustment of one rotor to the other, this technique can show improvements in mixing efficiency for some compound types.

Various rotor types and profiles are available for different mixing applications. The number of wings on the rotor, the shape of the wings, the position of the wings, the wing length, and the angle to the central axis of the rotor creates variations in rotor design. Correspondingly, this also affects the mixing effect of the rotors and the energy they introduce into the compound, both of which of course vary from one rotor design to another. The actual mixing effect can be defined as dispersive mixing and distributive mixing.

The gap between the two rotors creates a large area into which the compounding ingredients can be fed so that the feeding behavior of tangential rotors is typically very good.

Tangential mixers require a drive power of approx. 7 kW per liter of useful volume. This drive power can vary up or down, depending on the rotor type used and/or the kind of compounds being mixed.

However the mixer cannot be filled completely, because in order to achieve good mixing levels the compound needs free space in the mixing chamber so that the rotors can move it around in an optimal manner. This means that the actual useful volume is approximately 70 to 85% of the free volume of the machine. The ratio of useful volume to free volume is known as the fill factor (e.g., 70% = fill factor 0.7). The optimal fill factor for a particular compound depends on the compound itself (composition, kind of ingredients, hardness of compound, etc.) so that this range of values should be seen only as a broad reference guide.

Tangential mixers are manufactured with free volumes ranging from around 1 to 650 liter.

1.2.2 Intermeshing Rotors

In 1934, *R. Cooke* filed a British patent application [1] for an internal mixer with intermeshing counter-rotating rotors. It is an interesting coincidence that almost at the same date (just 4 months after Cooke's British patent application) *Werner & Pfleiderer* (Albert Lasch and Ernst Stomer) also filed a German patent for an intermeshing mixer [15]. However, manufacturing of such a machine only began in the 1950s by *Francis Shaw* and in the beginning of the 1980s by *Werner & Pfleiderer* [2]. In this design (Fig. 1.4), the paths described by the rotor tips actually overlap. The radial relationship between the rotors must therefore be permanently fixed and both rotors run at the same speed. Compared to the tangential system which exhibits an inner surface area of nearly the same size, and at a comparable rotor diameter, this design produces a reduced mixing chamber volume but provides a larger surface area to volume ratio to improve cooling. Temperature control of the compound is thus considerably more effective and higher energy input is possible at the same maximum mixing temperature. In addition, the markedly lower clearance between the two rotors in the intermeshing zone leads to a more vigorous mixing action. At the same time, the fill factors for the intermeshing system are approximately 5% smaller than those of the tangential system. The sizes of these mixers vary between 0.3 and 550 liter free volume and the drive power required is approx. 10–20% higher than for a tangential mixer of the equivalent size.

Figure 1.4 Sectional drawing of an internal mixer with intermeshing rotors

1.2.3 Technological Comparison

The relative properties of each rotor system are shown in Fig. 1.5. The advantages of the tangential mixer are its good feeding and discharge qualities and the high level of machine utilization. In contrast, the intermeshing machine primarily offers processing advantages and improvements in compound quality.

1 Internal Mixer – Configuration and Design

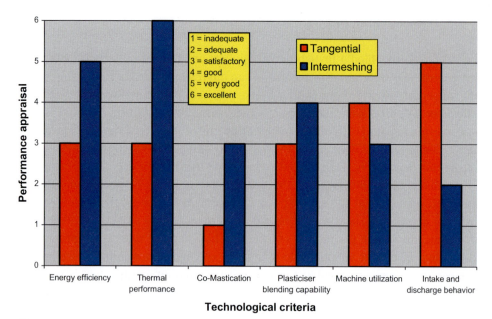

Figure 1.5 Technological comparison of tangential and intermeshing rotor systems

Figure 1.6 shows a comparison of the free volumes and internal surface areas for different tangential (N) and intermeshing (E) mixer sizes. With a free volume of approximately 255 liter, the GK 225 N has an internal surface area of 6 m². At the

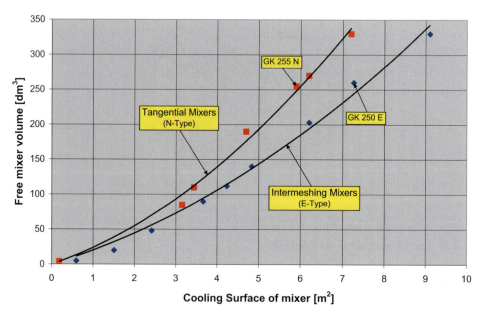

Figure 1.6 Comparison of free mixer volumes and cooling surfaces of tangential and intermeshing internal mixers

same time, the GK 250 E, with an almost identical free volume, has an internal surface area of 7.2 m², which is about 20% larger. Considering this, it is not surprising that intermeshing mixers have better cooling characteristics than tangential machines.

Traditionally, tangential mixers were used extensively in the tire industry. However, this has changed somewhat since the arrival of silica compounds. Mixing this type of compound calls for a defined and reproducible temperature history which is much more readily achieved with an intermeshing mixer because of its better cooling.

1.2.4 Intermeshing Rotors with Variable Clearance (VIC)

Compared to tangential mixers the small clearance between the two rotors in intermeshing mixers produces an additional mixing effect. But during the first phases of the mixing process, when raw materials are feed into the mixer, the small gaps are complicating the filling of the machine. The raw material has to be forced through the gap zone between the rotors and ram lowering takes a long time. The larger the gap between the rotors, the faster is the intake behavior of the mixer.

The above statement makes it evident that the possibility of varying the clearance between the rotors in an intermeshing type mixer provides the opportunity to influence the mixing process. At the beginning of the mixing procedure the rotors have to be open to allow a short intake phase, while later on the rotors will be closed to realize a good mixing effect. When the rotors are moved during the mixing process, it must be considered that the gap between the rotor tips and the chamber wall will be changed, too (Fig. 1.7). This affects the shear rate in the gap and the amount of material flowing through the tip zone.

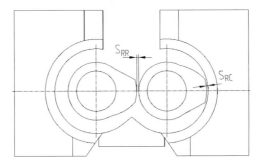

Figure 1.7 Variable intermeshing clearance of an intermeshing mixer (s_{RR} = clearance between rotors; s_{RC} = clearance between rotor and mixing chamber)

The movement of the rotors affects not only the intake behavior of the raw material, it also influences the mixing process itself (shear rates, movement of material, cooling efficiency), the temperature rise of the batch, and the energy consumption during a mixing cycle. These relationships are very complicated and not easy to handle. Additional experiments have to be carried out to gain experience with such a new mixing parameter.

1.2.5 Tandem Mixing

The principle of tandem mixing was invented by *J. Peter* in 1987 [3]. His idea was to combine two-stage mixing, in which the master batch and the final batch are mixed consecutively in the same mixer, into a one-stage process with two mixers. Tandem mixing uses a combination comprising a conventional internal mixer and a significantly larger tandem mixer without floating weight situated underneath of the first mixer. After premixing in the ram type mixer, the compound is dropped directly into the tandem mixer. Here, the compound temperature is reduced to between 100 and 120 °C before the reactive substances are added. At the same time, another master-batch is being prepared in the ram-type master-batch mixer. From the tandem mixer the final batch is dumped on a mill or into an extruder where it will be sheeted and conveyed to a batch-off unit for further cooling. The main advantage of this process is that the master-batch does not need to be stored between first and second stage. Transport and labor costs can be reduced and the compound quality is no longer depending on storage time. More than 15 years ago,

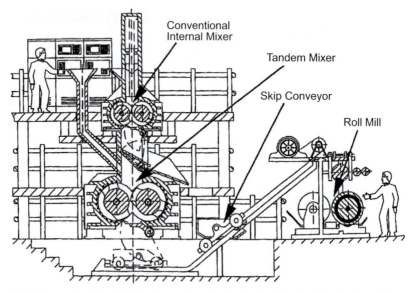

Figure 1.8 Schematic of the first tandem mixer line in the technical rubber industry [3]

a tandem mixer line was realized in a technical rubber factory where it is still in operation (Fig. 1.8). However, the general concept was not accepted by the market at that time. With the big advantages of the intermeshing systems, technical rubber compounds can also be mixed in one stage.

At the begin of the new millennium, with the intensive use of silica technology, the tandem mixer was born again. At the beginning of the silica technology in the tire industry, silica compounds had to be produced on already installed mainly tangential mixers. These types of compounds are totally different from conventional compounds with carbon black as the main filler. Here, a chemical reaction, the so called silanization, occurs inside the compound during mixing of silica compounds. So the internal mixer was not only a machine for mixing, it also became a chemical reactor. This additional function of the mixer increased the mixing time, so that silica compounds had to be mixed in several more stages than conventional black compounds. At that time, the tandem technology was introduced for silica mixing. The process was split into a first mixing stage, which was done in an intermeshing mixer with floating weight and a second reaction stage, which was done in a tandem mixer of intermeshing type below (Fig. 1.9).

The advantage of tandem mixing for silica compounds is that in both machines mixing of the compound can be adapted to its unique process requirements. In the upper machine, good filler dispersion can be achieved with a high energy input. The ram is used to force the material between the rotors, where it is subjected to

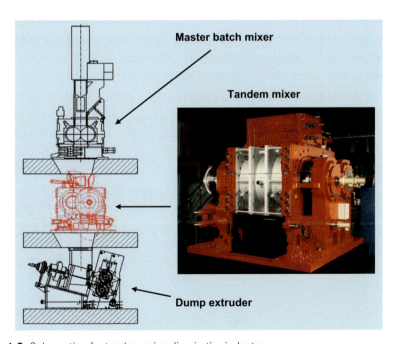

Figure 1.9 Schematic of a tandem mixer line in tire industry

high elongation forces. These forces are responsible for the excellent mixing efficiency in an intermeshing machine. A high specific energy input quickly increases the temperature of the compound. Once the compound reaches a certain temperature, it is dropped into the tandem machine. Directly after this, the next mixing cycle is restarted in the upper machine. At the same time, the previous batch is finished in the tandem machine. Tandem mixing lines can have up to double the throughput of a conventional mixing line.

The tandem mixer is considerably bigger than the upper machine, so that the mixing process will take place in an under-filled mixer. This is absolutely essential because of the relative poor intake behavior of the machine. Otherwise, the batch would not pass the small gap region between the rotors. Compared to a conventional mixer of the same size, a tandem machine requires a lower specific drive power.

The batch volumes in the upper machine and in the tandem mixer are the same, because no additional raw material will be fed into the tandem mixer. Compared to the smaller upper machine, a tandem mixer has a bigger mass and offers more cooling area to the batch. Therefore, the batch temperature can be controlled very efficiently. A special control system regulates the rotor speed, so that the batch temperature can be kept within a given range. In this case, the energy input over the drive (mechanical energy) and the energy output over the cooled mixer parts (heat energy) are balanced.

When a chemical reaction takes place during the silica mixing process, alcohol is produced as a reaction product. The under-filled tandem mixer is well equipped to vent this vaporized alcohol because of the big free volume inside the machine and the open feeding entrance (no ram). With additional venting points above and under the tandem mixer, the produced vapor is vented out of the machine.

Relying on the reaction kinetic of the silanization process, the mixing procedure can be controlled. After silanization is finished, the batch is dropped out. Compared to conventional mixing of rubber compounds, neither temperature nor energy input are the criteria for to drop out the batch, but rather the end of the chemical reaction is the drop criterion.

Because mixing is performed in parallel in the upper and lower units, the mixing time for production of silica compounds can be reduced significantly using tandem technology. Therefore, it is possible to achieve a considerably higher production capacity together with high compound quality.

1.3 Feeding Hopper

1.3.1 Design Considerations

The feeding hopper of an internal mixer consists of an upright rectangular housing inside of which is a ram. The latter can be moved up or down vertically by a pneumatic cylinder situated centrally above the feeding hopper or by one or more hydraulic cylinders. During the mixing process, the solid ingredients of the rubber compound, and occasionally liquid ingredients, are fed into the mixing chamber via the feeding hopper. Several openings are provided in the sides of the feeding hopper for this purpose.

A feeding door is located at the front of the feeding hopper and is opened and closed by a hydraulic cylinder. Polymers and the small chemicals are normally fed into the mixer through this door. In some cases, mostly in older installations, bulk fillers are also fed here. The duel task of the feeding door is to provide access to the mixing chamber when the door is open and compound ingredients are being fed in and to seal off the chamber when the door is closed to prevent dust from escaping.

Openings in the back and in the sidewalls of the feeding hopper are used to feed bulk fillers and small chemicals automatically from upstream materials weighing and feeding systems. For maintenance routines or cleaning work, the feeding hopper is provided with openings fitted with a safety interlocked cover or door to allow access to the interior.

A "cooling chamber" inside the ram body allows independent temperature control. The ram temperature can be set to prevent hot ram surfaces during mixing and thus to avoid compound sticking to it.

At the start of the mixing process, powder fillers are fed into the mixing chamber via the feeding hopper and this produces large amounts of dust inside the mixer. When the ram moves down, air charged with dust flows upwards from the mixing chamber through the gap between ram and hopper into the body of the feeding hopper. This dust-loaded air can be exhausted via an opening placed above the feeding door.

To prevent dust settling on the top of ram, its upper surfaces are all angled. Dust slides down and drops back into the mixing chamber through the gaps between ram body and feeding hopper walls. Ram cleaning can be improved further by the use of air nozzles in the feeding hopper, which blow air onto the ram in the raised position.

During the mixing process, the ram (Fig. 1.10) presses the ingredients and/or the compound into the mixing chamber so that they all participate in the mixing

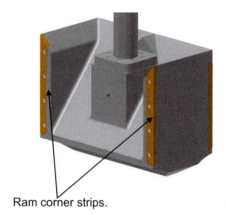

Ram corner strips.

Figure 1.10 Ram (floating weight)

process. In addition to the pneumatic or hydraulic pistons/cylinders, the ram is guided by a set of 4 replaceable wear ledges attached to the vertical corners of the ram. These have to be exchanged when excessive wear becomes noticable; otherwise the ram scratches the hopper walls.

1.3.2 Pneumatic Feeding Hopper

For historic reasons, pneumatic feeding hoppers (Fig. 1.11) were the first to be used. The ram is connected through a ramrod to the piston of a pneumatic cylinder. As the ram moves, air is also exhausted from the cylinder into the atmosphere, resulting in high noise levels during operation. Therefore, large exhaust air silencers are normally required.

Figure 1.11 Pneumatic hopper assembly

The ram has two external cooling pipes through which cooling water can flow through the ram body. These pipes and the ramrod all pass through static seals in the top of the feeding hopper to prevent the escape of dust.

Ram movement and the pressure applied to the compound by the ram depend on the pressure available in the compressed air supply network. Fluctuations of pressure in these networks are commonplace, due to varying consumption demands from other users or day/night operations. Pressure changes in the air network affect the consistency of the ram pressure, which in turn affects the mixing process and thus the quality of the compounds produced. In order to avoid this problem and the high costs of compressed air production/distribution, the hydraulic feeding unit was developed.

1.3.3 Hydraulic Feeding Hopper

Hydraulic feeding units (Fig. 1.12) have proven themselves in practical use in recent years. Due to their economic benefits they have replaced pneumatic feeding units to a large extent. In the hydraulic hopper with two cylinders, a crossbar is used to transfer the movement of the cylinders to the ramrod. The ramrod is fixed to the crossbar by a flexible mounting and two additional guide columns ensure the precise movement of the crossbar and ram. The external position of the cylinders ensures that hydraulic oil cannot contaminate the compound via the feeding unit. Ram cooling is affected through the center of the ramrod without the need for two separate pipes required for pneumatic hoppers.

Two-cylinder-design   Single-cylinder-design

Figure 1.12 Hydraulic hopper assembly – different designs

With the recently developed single-cylinder design, the ram is connected through a ramrod directly to the piston of a hydraulic cylinder in a similar way as the construction of a pneumatic feeding hopper. In addition, special seals are used to prevent the possibility of hydraulic oil contaminating the compound. The ram cooling pipes are also similar to those used on a pneumatic hopper.

1.3.4 Comparative Aspects

Hydraulic feeding hoppers offer several technical advantages compared to pneumatic feeding hoppers (Fig. 1.13):

Low Noise Levels
The high noise levels created when the pneumatic cylinder vents to the atmosphere can be avoided with hydraulic feeding hoppers.

Rapid Ram Movement
Very fast ram movements are possible with hydraulic hoppers. Complete ram strokes only take a few seconds and various speeds can be set within the stroke length to provide rapid speed and end of stroke cushioning.

Precise Ram Force Adjustment
The ram force can be set precisely with a differential hydraulic pressure regulator. Hydraulic pressures on both sides of the piston are measured and the forces calculated in both directions using the relevant effective piston areas. The difference between these forces is the actual ram force applied to the compound.

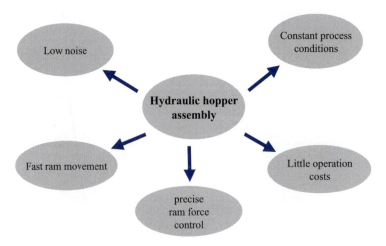

Figure 1.13 Advantages of hydraulic hoppers

Reproducible Process Conditions

Everyone with practical mixing experience knows that pressure fluctuations regularly occur in compressed air networks, due to varying user requirements. Within a mixing cycle, these can result in differences in the movement of the ram and in the force applied by the ram to the compound.

The hydraulic feeding hopper ensures that both the ram movement and the ram force are repeatable from batch to batch. This results in repeatable mixing conditions and thus significantly raises the quality of the compounds produced.

Reduced Operating Costs

A comparison of operating costs (Fig. 1.14) reveals significant reductions in energy costs when using a hydraulic feeding hopper. With the operating conditions and the costs of 0.014 € per m_N^3 for compressed air and 0.08 €/kWh for electrical power, as indicated in Table 1.1, the annual operating costs for a pneumatic feeding hopper for a GK135 E are approx 20,500 €. In contrast, the operating cost of a hydraulic feeding hopper (single cylinder) over the same period amounts to only 3,500 €, resulting in savings of 17,000 € per year in energy costs alone.

As a basis for the comparison, the same number of operating hours per year (6000 h/a) and ram pressure (6 bar) were selected. The calculations for the two intermeshing mixers, the GK 320 E and GK the 135 E, were based on an application in the technical rubber goods industry. The calculation for the tangential mixer (GK 255 N) was based on the costs for a final batch mixer in the tire industry.

The cost for compressed air (0.014 €/m_N^3) not only contain the cost of compression, but also includes additional costs such as assets, service, building, maintenance, cleaning, and spare parts among others.

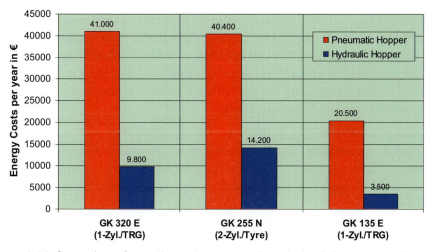

Figure 1.14 Comparison of operating costs – pneumatic vs. hydraulic hopper assembly

Table 1.1 Operating Conditions

Machine	Design	Batch quantity /hr	Operating hours / annualy	Double strokes / batch	Pressure on the compound
GK 320 E	Single cylinder	12	6000	5	6 bar
GK 255 N	Two cylinder	30	6000	3	6 bar
GK 135 E	Single cylinder	12	6000	5	6 bar

Single-cylinder and two-cylinder hoppers behave differently with regard to the energy they consume because of their different design. In the comparative calculations of operating costs, the single-cylinder design always performs substantially better than the two-cylinder design.

Changes in the basic data used for the calculation of production costs obviously lead to noticeable changes in the actual costs calculated. Comparisons for a specific case should therefore be made using local operating conditions and costs for compressed air and electrical power.

■ 1.4 Digital Ram Position Control

In standard internal mixers, ram speed and ram pressure are adjusted during the installation phase of the machine and they normally are not changed by the user of the machine.

During the production of rubber compounds in an internal mixer, different kinds of hopper problems can occur. Dusty fillers may come up over the ram and settle on top of the ram so that later batches can be contaminated by those fillers falling down from the ram into the chamber. Especially soft and sticky compounds tend to stick to the hopper walls and to sides and the backside of the ram. These phenomena can affect the functionality of the hopper assembly and will also increase wear. In an extreme case, the floating weight can stick inside the hopper and may not be able to move any more ("ram-jamming").

With a digital ram position control the compound quality can be improved and operation disruptions, as described above, can be avoided. Such a system consists of a displacement transducer which is installed above the hopper and a special control system. Depending on ram position, ram speed and ram pressure can be controlled (Fig. 1.15).

After an initial high ram speed (phase I in Fig. 1.15) at the beginning of the ram down phase, the ram speed was reduced during the end phase of the ram down movement into the neck of the mixing chamber (phase II in Fig. 1.15) and the raw

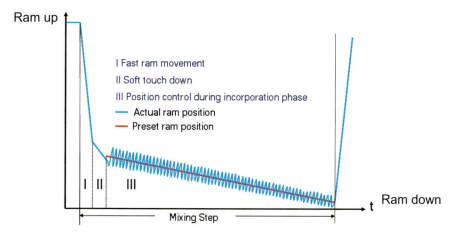

Figure 1.15 Ram movement controlled by ram position control system

materials were pushed into the mixing chamber on a preset ram track (phase III in Fig. 1.15). Using this type of ram control, the amount of filler forced on top of the ram can be reduced significantly. Therefore, the quality of the mix can be considerably improved and the mixing cycle can also be reduced (because of shorter cleaning steps during the cycle).

Another example of how ram position control can help to improve the mixing process and compound quality is shown in Fig. 1.16. During the production of a rubber compound using the upside down method, a considerable amount of free filler was blown on top of the ram. By reducing the ram speed, the contamination of the ram is efficiently reduced (Fig. 1.16).

To prevent too much material from being pressurized into the gap between ram and hopper walls, the final ram position at the end of the incorporation phase can be reduced several mm ("set back" S; Fig. 1.17). Thus, the compound can expand somewhat into the free volume under the ram when it is forced from the chamber

a) without ram position control   b) with ram position control

Figure 1.16 Highly (a) and little (b) contaminated ram

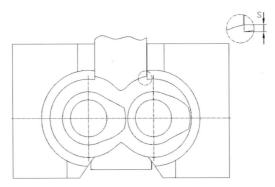

Figure 1.17 Ram end position controlled by ram position control system

area into the ram area by the wings of the rotors. The pressure under the ram is lowered and the tendency of the material to climb up into the gap is reduced.

A digital ram movement control system in an internal mixer improves rubber compound mixing by providing/ensuring:

- better compound quality
- less operation disruptions
- higher productivity
- less wear

1.5 Mixing Chamber

To maximize the service life of the mixing chamber, it must be of very solid construction. The contact surfaces with the compound should be hard, tough, and corrosion-resistant in order to withstand its continual abrasive action.

The cooling ability of the mixer is also of fundamental importance. All parts of the mixer that come into contact with the compound should be provided with a temperature control unit. Although the rotor end plates, for example, make only a very minor contribution to the thermal efficiency of the machine, parts of the mixer that do not have temperature control may cause overheated areas to which the compound may adhere causing discharge problems. Effective temperature control over the entire mixer means good dissipation of heat from the compound and lower batch temperatures.

1.5.1 Hard-Coating

All inner surfaces of the mixing chamber that come into contact with the compound should be provided with wear protection. Normally this means a hard steel coating welded on the base material of the mixer parts (e. g., ram, chamber, rotor, rotor end plates, drop door) in order to withstand high abrasion and corrosion caused by the materials to be mixed.

Figure 1.18 shows a section through a mixing chamber with hard-coating on the chamber and rotor surfaces.

The first hard coatings used in internal mixers were made from cobalt based alloys with the brand "Stellite®". These are hard alloys containing a high amount of chromium, which form cracks during the hard coating process because of their composition. A newer material called WP 53 M was developed in the beginning of the 1990s . With this new alloy, crack-free hard coatings are available.

Figure 1.19 shows two photomicrographs of different hard coatings, Stellite 1® and WP 53 M. The crack in the Stellite 1® coating can be seen clearly, whereas no cracks can be seen in the WP 53 M coating.

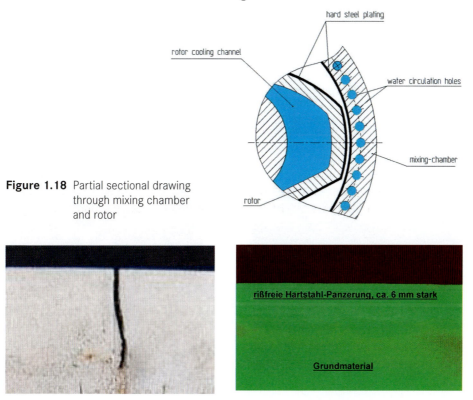

Figure 1.18 Partial sectional drawing through mixing chamber and rotor

Stellite 1　　　　　　　　　　　WP 53 M

Figure 1.19 Hard coatings: Stellite 1 and WP 53 M (polished micrograph sections)

When mixing chemically aggressive compounds or compounds with high levels of humidity, cracks in the hard coating allow corrosion to occur in the base metal of the hard coated item. Thus, the adhesion of the hard coating to the base metal is weakened and pieces of the hard coating can break off.

Figure 1.20 shows the operational surface condition of a typical Stellite® hard coating and the efflorescence of corrosive products on the hard coating after a longer period of time in a machine out of operation. Due to their larger volume, the oxides formed as part of the corrosion process permeate through the crack up to the surface of the hard coating.

The main properties of the new hard coating material are summarized in Fig. 1.21.

Due to automated application, the hard coating has a constant thickness and a homogeneous structure. The combination of the physical properties of hardness and tensile strength ensures very good abrasion resistance and because of its high chromium content the new hard coating is also corrosion resistant.

However, the major advantage is the absence of cracks. Crack free hard coatings effectively prevent the undermining corrosion that occurs with standard hard coatings which develop cracks. It is therefore possible to reduce the distance between the surface of the hard coating and the cooling channels beneath it, which results in a considerable improvement in the cooling ability of the internal mixer. The crack free hard coating material has now been in service for more than 15 years and, depending on the specific application, a service life ranging from 5 to 15 years can be expected.

For very abrasive and/or corrosive compounds, specially adapted hard coatings have to be used. Because of their high abrasive and corrosive behavior, silica compounds used in the tire industry require special (dual) protection systems.

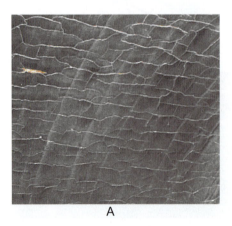
A

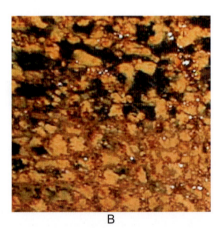
B

Figure 1.20 Hard coating with cracks in a running machine (A) and efflorescence of corrosive products in a machine out of work (B) over a longer time period

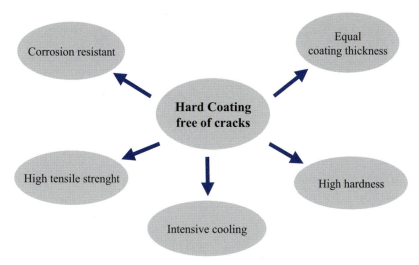

Figure 1.21 Properties of WP 53 M crack free hard coating material

1.5.2 Dust Sealing

At the start of the mixing process, the mixing chamber is charged with fillers and other powdery ingredients. These fillers have small particle sizes and can pass through the annular gap between the rotor shafts and the rotor end plates/mixer end frames, to escape as dust to the outside of the mixer. In addition to this dust, liquids, pastes, and solid materials can also escape from the mixer at the same points as the mixing process continues.

The two rotor shafts pass through the end of the mixing chamber at both ends, which means that there are four annular gaps that need to be sealed. Each seal is made mechanically, using two rings that slide on each other (Fig. 1.22). One ring is fixed onto the rotor shaft and turns with the rotor. The other ring is fixed to the mixer end frame and remains static. Both rings are made in a split design and can therefore be removed and refitted without removing the rotors from the mixing chamber.

To ensure a long service life, the contact surfaces on the two rings are hard-coated and ground. During operation, they are lubricated with oil fed to them via several drillings in the static ring to reduce friction.

In addition, the ring gap between the rotor shaft and the rotor end plate/mixer end frame are also fed with process oil via several separate drillings. This process oil binds the dusty compound ingredients into a paste in the ring gap. When the mixer idles between mixing cycles, the process oil tends to clean out the ring gap.

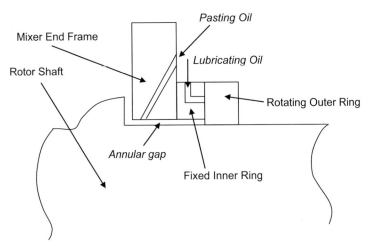

Figure 1.22 Operating principle of dust seal

Types of dust seals:
- Self-sealing and adjusting dust seals   (SSA)[1]
- Spring loaded dust seals   (GA)[2]
- Hydraulic dust seals (yoke)   (YH)[3]
- Hydraulic dust stop (cylinder)   (CH)[4]

In the case of self-sealing dust seals, the pressure of the compound in the mixer generates the contact pressure between the two rings. During the mixing process the pressure on the seal is high and between mixing cycles and during periods of inactivity it is low; the system is in practice left to run itself. The self-sealing dust seal is only used in very old mixers. Only the spring loaded and hydraulic dust seals are now used in modern machines.

1.5.3 Spring Loaded Dust Seals

In the case of spring-loaded dust seals, the rotating outer ring on the rotor shaft is pressed onto the static inner ring, fixed in the mixer end frame by several spring packs (8 to 12 packs). These are fitted around the rotating outer ring (Fig. 1.23). The ring contact pressure can be adjusted by varying the preloading on these spring assemblies.

Process oil is pumped into the annular gap in front of the dust seal in order to make a paste from the dusty fillers while lubricating oil is pumped onto the contact surface between the two rings to reduce friction. Two separate lubrication pumps are used.

[1] SSA = Self-sealing adjusting; [2] GA = general adjusting; [3] YH = yoke hydraulic; [4] CH = cylinder hydraulic)

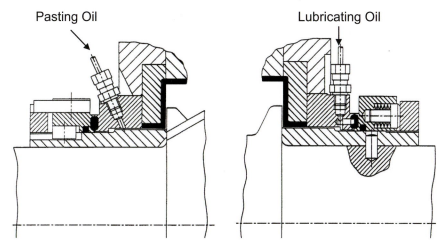

Figure 1.23 Spring loaded dust seal (GA – dust seal; sectional drawing)

In order to achieve pasting, the process oil flow can be adjusted with a variable two-speed or even frequency-controlled pump. An empty mixer normally operates with 50% less process oil flow rate than a full mixer, so that during idling time excess oil does not flow into the mixing chamber. The pump supplying lubrication oil to the rings' contact surfaces maintains a constant delivery. Apart from varying the speed of the pumps, oil flow through each individual feed pipe can be controlled separately via manual valves attached to the pumps.

Improved adjustment of oil flow can be obtained using frequency controlled lubrication pump motors. Here, the motor speeds are infinitely variable and the oil quantities can be regulated by the mixer control system. This allows even better adjustment of the lubrication system to match the mixing process and the types of compound being produced. An optimal sealing effect, with minimum oil consumption, can be ensured by automatic intelligent control of the lubricating pumps.

Because of the abrasive nature of the pastes formed in the annular gap, the metal surfaces in this area are provided with special wear-protection.

1.5.4 Hydraulic Dust Seals with Yoke

With hydraulic dust seals the separation plane itself fits closer to the rotor body (Fig. 1.24) and is not visible from the outside of the mixer. Again, two lubricated rings slide against each other to produce the seal. In this case, the outer static ring is fixed in a yoke and pressed by a rocker arm against the inner rotating ring, which is fixed to the rotor shaft (Fig. 1.25). The contact pressure is applied using a hydraulic cylinder located at the end of the arm. The contact pressure can be varied by adjusting the hydraulic pressure in the cylinder from zero to maximum.

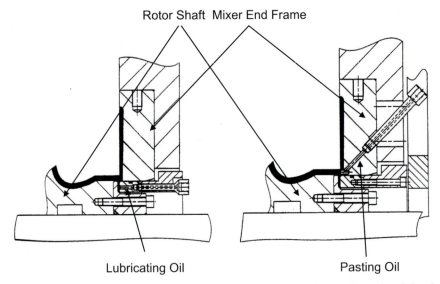

Figure 1.24 Hydraulic dust seal (WYH = "Werner" hydraulic yoke dust seal; sectional drawing)

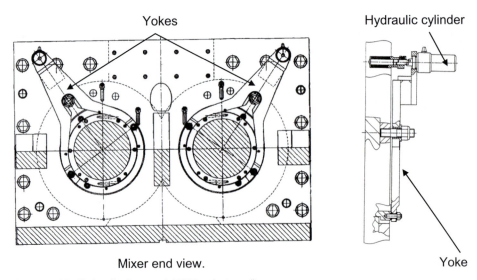

Figure 1.25 Hydraulic dust seal (WYH – dust seal)

Lubricating oil for the ring contact surfaces is fed through the outer static ring. Process oil for pasting is pumped into the pasting area through drillings in the mixer end frame and rotor end plate.

1.5.5 Hydraulic Dust Seals with Cylinders (CH)

With the hydraulic dust stop with a yoke, the fixed ring is pressed against the rotating ring over two points of force transmission. To provide the sealing rings with a more equal force distribution, the fixed ring is equipped with four hydraulic cylinders (Fig. 1.26). They are installed directly behind the ring. In this case, there are two more transmission points of force which press the rings together. The lubrication of the WCH dust stop works according to the same principle as the WYH-dust stop.

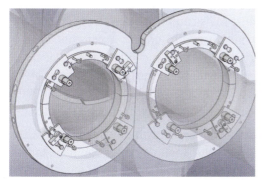

WCH dust stop, each with 4 hydraulic cylinders   One element with hydraulic cylinder

Figure 1.26 Hydraulic dust seal with cylinders (WCH "Werner cylindrical hydraulic" – dust seal)

1.5.6 Comparison of Spring Loaded and Hydraulic Dust Seal Systems

Table 1.2 shows a comparison of the spring loaded and hydraulic dust seal systems. In the case of the GA dust seal, the spring packs are distributed around the circumference of the ring and the force is applied more evenly compared to the hydraulic system. Here, due to the effect of the yoke, force is applied in only two spots, one at each side of the ring. With spring loaded dust seals, the contact pressure can be set individually at each of the four dust seals on the mixer. Setting the

Table 1.2 Comparison between Spring Loaded and Hydraulic Dust Seals

Type	Force introduction	Individual adjustment	Process adjustment	Accessibility	Inspection
GA	+++	+++	–	+++	++
WYH	+	–	+++	+	+
WCH	++	–	+++	+	+

+++ excellent   ++ very good   + good   ○ satisfactory   – not possible contact pressure individually can compensate for different loads on the seals, due to the movement of the compound within the mixing chamber and axial movement due to thermal expansion of the rotors.

Power to the hydraulic cylinders fitted to the hydraulic dust seal is supplied from a single power source, which means that the contact pressure is identical on all four seals. The main advantage of the hydraulic dust seal is that it is possible to adjust the contact pressure during the course of the mixing process or to the type of compound being mixed. Accordingly, the plant operator can react to individual requirements for compounds that require different types of processing. Thus, wear of the rings and the consumption of lubricants can be reduced to a minimum.

For maintenance work, spring loaded dust seals can be accessed from the outside and wear on the rings can be easily measured. To ensure a constant contact pressure, the adjustment of the GA dust seal should be checked at regular intervals. Replacement of the rings is equally manageable for both kinds of seal. To replace the rings on all four of the dust seals on a mixer normally takes one fitter approximately one to two days, depending on the accessibility around the mixer.

From a process point of view, the major advantage of hydraulic dust seals is the elimination of the annular gap between the rotor shaft and rotor end plate/mixer end frame. Deposits of material in the annular gap and the potential contamination of the compound by these deposits are no longer possible. Hydraulic dust seals are therefore recommended for the production of non-black or colored mixes and compounds of very sensitive nature.

1.6 Temperature Sensor

The temperature of the rubber compound is recorded during the mixing process by temperature sensors that extend into the mixing chamber. The signals from these sensors are routed into the mixer control system, where they serve as a step criterion for the mixing process. Temperature sensors can be positioned in the end frame of the internal mixer and also in the discharge door (Fig. 1.27).

Temperature recording within the mixing chamber is affected by the following conditions:

Compound Flow around the Temperature Sensor

A good contact of the compound to the temperature sensor, enforced by the flow pattern in the mixer, is a basic requirement for temperature measurement in the mixing chamber of an internal mixer. To ensure this, the sensor has to extend deeply enough into the mixing chamber so that the material being moved by the

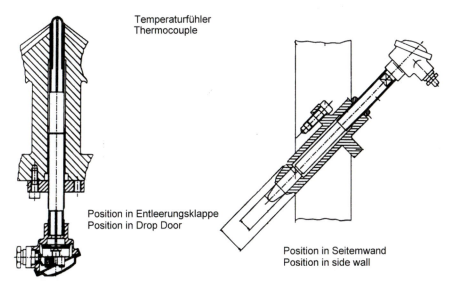

Figure 1.27 Typical installations of temperature probes in internal mixers

wings of the rotors can flow around the sensor tip. In the construction of modern mixers the temperature probe is positioned much deeper into the mixing chamber than before. The rotor end plates, which effectively form the end of the mixing chamber, now have an additional recess that makes this deep sensor position possible.

The flow around the temperature sensor depends on the machine type (tangential or intermeshing), the geometry of the rotors (full-4-wing or ZZ 2, etc), the position of the sensor inside the mixing chamber (end frame or discharge door), the type of compound (soft or hard), and the mixing process (high or low fill factor). An example will show how the temperature measurement can be influenced. In Fig. 1.28 two temperatures recorded over a mixing cycle are shown. One signal originates from a sensor installed in the discharge door and the other was measured

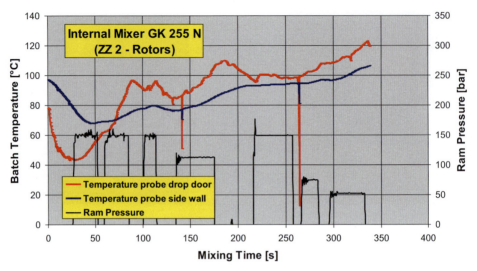

Figure 1.28 Temperature profiles from a temperature sensor in the mixer end frame (blue line) and in the discharge door (red line)

by a sensor in the end frame. While the sensor in the discharge door closely follows the mixing process, the sensor in the end frame almost always indicates a lower temperature and its reaction is very slow. This is due to unsatisfactory compound flow around the temperature sensor tip caused by the normal mixing action of the outside wings of the ZZ 2 rotors, which move the compound from the ends of the mixing chamber towards the center. Thus, insufficient material remains at the chamber ends, unable to ensure a continuous contact with the temperature sensor. For the ZZ 2 rotor geometry it is well known that the compound flow around temperature sensors in the mixer end frames is typically unsatisfactory so that temperature probes have to be installed in the drop door. This factor should be considered carefully, especially with older machines where existing rotors (for example 2-wing type) are exchanged against ZZ 2 rotors.

Sensitivity of Temperature Sensors

By constant advancement in the design of temperature sensors (construction methods and modern materials) their sensitivity has been continuously improved. In Fig. 1.29 the dynamic behavior of a conventional temperature sensor is compared to that of a newly developed one. The tips of both sensors were dipped into a temperature controlled oil bath at 150 °C and their temperature gradients recorded. The temperature curves produced show that the new sensor exhibits a substantially better reaction behavior than the conventional type. The new sensor reaches the oil bath temperature of 150 °C after approximately 50 seconds, whereas the conventional sensor needed twice that time.

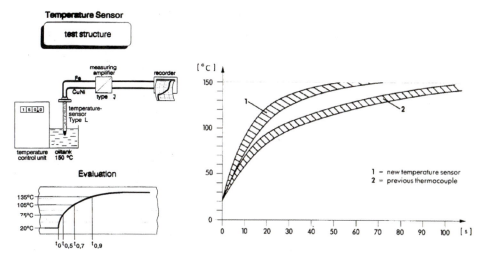

Figure 1.29 Temperature sensor – test method showing dynamic behavior

By the selection of a suitable combination of materials in the construction of the new temperature sensor, heat dissipation from the sensor tip into the sensor body was reduced substantially and its reaction time reduced by half.

1.7 Plasticizer Oil Injection

By injecting plasticizers it is possible to feed liquid components into the mixing chamber during the mixing process. Injection valves are fitted into the mixing chamber halves (Fig. 1.30). In smaller mixers one valve can be installed in each mixing chamber half. Bigger machines have two holes in each chamber half

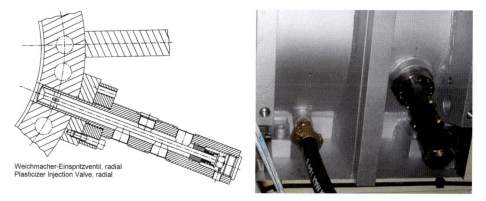

Figure 1.30 Injection valve radially fitted into the mixing chamber

prepared for installation of injection valves. Combinations of these four positions not only permit the injection of large quantities of plasticizer into the compound in a short period of time, but also the injection of different types of plasticizer via the various valves.

The plasticizer valve consists of a housing with several external pipe connections. Inside the housing is a spring-loaded valve plunger. The plunger moves forward against the spring once the plasticizer oil pressure reaches a sufficient level to overcome it. This then frees the valve opening and the plasticizer oil is injected into the mixing chamber. When injection is complete, the valve is returned to a closed position by the action of the spring.

When two valves are installed in a machine, one valve should be positioned on the water side and the other on the drive side. This makes sure that the plasticizer will be optimally distributed inside the compound.

If big amounts of plasticizer are injected in an internal mixer, the temperature measurement can be influenced. The oil acts like a lubricant and prohibits intensive contact between compound and the tip of the temperature sensor. In this case, the oil should be split up into smaller amounts which are injected consecutively or by controlled oil injection.

1.8 Rotors

The mixing effect in an internal mixer is determined to a large extent by the type and mode of operation of its rotors. The rotors have three functions:
- Impart shear and elongation into the compound (dispersive mixing)
- Distribute ingredients inside the compound (distributive mixing)
- Dissipate heat out of the compound

1.8.1 Assembly and Cooling

In the design of mixer rotors, a distinction is drawn between one-piece and two-piece rotor construction (Fig. 1.31). One-piece rotors are made from a single casting that forms the rotor body and the two shafts. They must be extremely sturdy in order to withstand the forces acting on the rotor during the mixing process. Their strength is achieved by rotor body walls of considerable thickness. Cooling is effected through a pipe with several nozzles situated inside the rotor body (spray cooling). The cooling water is sprayed upwards, against the upper inside wall of the rotor body and then flows back into the lower part, where it flows

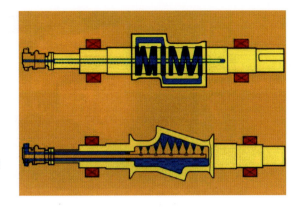

Figure 1.31 Two-piece rotors (top) and one piece rotor (bottom)

out through one of the two rotor shafts. Basically, there is effective cooling in the top half of the rotor body only, because the lower half is filled with water and the cooling effect in this area is very poor.

With the two-piece rotor construction, the cast rotor body is made independently of the rotor shaft and secured to it by shrink fitting and keying. Cooling water flows into the rotor through a rotary joint and a pipe which is installed in the open shaft. Inside the rotor the water flow follows a spiral channel in the cast rotor body. Through an annular channel between pipe and shaft the water can flow out again. Compared to spray cooling, the cooling system inside the rotor body provides a forced water flow that produces a considerably better cooling effect inside the complete rotor.

Another major advantage of the two-piece rotor is the reduced wall thickness between the outer rotor surface and the cooling channels below it. This ensures a minimum distance between the compound and cooling medium, which again vastly improves heat transfer from the compound.

Because two piece rotors consist of a very stabile shaft in their center, they offer a much bigger resisting torque. Therefore, these rotors are much stronger and the risk of damage caused by an overload is lower.

1.8.2 Rotor Bearings

Each rotor in an internal mixer is mounted on two self-aligning roller bearings (Fig. 1.32) that cater to both radial and axial loads. For larger mixer sizes an additional thrust bearing is fitted to each rotor (Fig. 1.33) to take up the increased axial loads imposed on the rotors during the mixing process. This ensures even under high axial loading only small movement of the rotors and thus small gaps between rotor and rotor end plate.

Figure 1.32 Spherical roller bearing installed on a rotor

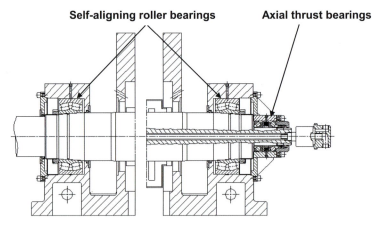

Figure 1.33 Bearings inside an internal mixer (sectional drawing)

1.8.3 Rotors for Tangential Internal Mixers

Basics

In tangential machines two rotors are placed parallel to each other, working in a counter-rotating way. Each of those rotors has a cylindrical core on which several wings (or vanes) are positioned. Rotors in tangential mixers traditionally used to run at different speeds. The rotor speed of the faster running rotor is about 10 to 15 % higher. Lately, tangential rotors running at the same speed can provide significant improvements, if the two rotors are positioned in the right configuration to each other.

The mixing process is mainly influenced by the nature of the rotors. Therefore, the rotors were always the focus of developers of internal mixers. Distinctive features of tangential rotors are:

- Number of wings
- Position of wings
- Length of wings
- Angularity of wings
- Outline of wings (active and passive side)
- Tip width of wings

With these parameters, a variety of different designs are possible. A historic survey of the tangential rotor development is given in the following.

Two-Wing Rotors

Modern internal mixers were developed in the beginning of the 20th century after the inventions of F. H. Banbury [4, 5]. At that time, rotors equipped with 2 wings each (Fig. 1.34) were used. One of the wings extends axially for a substantially greater distance than the other wing. The wings are diametrically located relative to each other at their plane of juncture, with the long and short wings helically disposed in opposite directions relative to the axis of the rotor.

Technically, this type of rotor is universally applicable and has therefore a very broad area of applications, from general rubber compounds to brake linings, plastic mixing, and other special applications. Today however, 2-wing rotors are used only in very old machines because the productivity with such type of rotor is not very high.

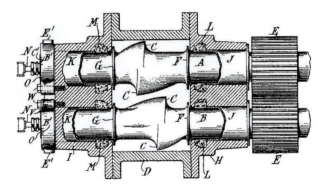

Figure 1.34 Internal mixer with 2-wing rotors [6]

Four-Wing Rotors

Because of the two additional wings, rotors with 4 wings have a bigger volume than those with 2 wings. The free volume inside a 270 l mixer (compared to a mixer using two-wing rotors) decreases by approx. 5% with the use of 4-wing rotors. This disadvantage is compensated more than adequately with better mixing performance. Four-wing rotors are mixing and dispersing carbon black more rapidly than two-wing rotors, however, they introduce considerably more heat into the mix. Compound temperature rises faster and higher drop temperatures are reached. The productivity of high intensity internal mixing machines equipped with 4-wing rotors is much higher. Because of their effectiveness, four-wing-rotors have been achieved first in tyre industry and later also in technical rubber industry. As follows some important types of 4-wing rotors are introduced.

N-Rotor (Normal Rotor; also Called Standard-Rotor)

With the beginning of the synthetic rubber era in the late 1930s and the starting 1940s, the N geometry rotor (Fig. 1.35) was developed by *Lasch and Frei* [7]. Basically, it is in a 2-wing rotor with 2 long wings, one from each side of the rotor, and two additional short wings in the center. The height of these two small wings is only 70% of the bigger ones. The N-rotor was replaced gradually in the 1960s and the 1970s by the next generation of new 4-wing rotor types. Now N-rotors are used in most cases in older machines or for special applications.

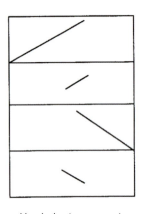
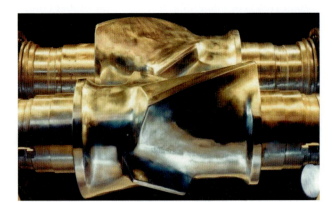

Unwind rotor geometry   Pair of N rotors

Figure 1.35 N-rotor

Full-4-Wing (F-4-W)-Rotor

Tyson and Comper [8] of the *Goodyear Tire and Rubber Company* invented the first full four wing rotor (abbr.: F-4-W), which was used worldwide in tire mixers (Fig. 1.36). The two longer wings are on the same end of each rotor, and the shorter flights are placed on the opposite side (Fig. 1.37).

The very long main wings, with their relatively low pitch angle and large flight depth, lead to a very rapid material intake. In addition, it is also possible to impart

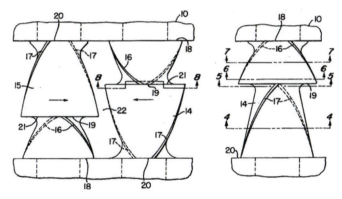

Figure 1.36 Full-4-wing rotor invented by Tyson and Comper [8]

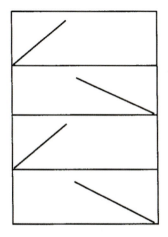

Unwind rotor geometry

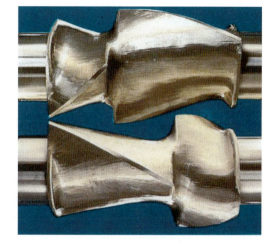

Pair of full-4-wing rotors

Figure 1.37 Full-4-wing rotor

large amounts of specific energy in a short period of time. These particular characteristics and good discharge properties have predestined this type of rotor for use in the tire industry, where it is in common use, in particular for master batch mixers.

ST®-Rotor (Synchronous Technology)

After a long period with only little further development in rotor technology, an improved rotor was patented by *Nortey* in 1987 [9]. While in the 4-wing rotor the two long wings and the two short wings are located on each side of the rotor, this is different in the ST rotor. Here, each long wing starts at another end of each rotor – the same principle is applied to the short wings (Fig. 1.38). The helix angles of the two wings on each side are the same, but different from one side to the other.

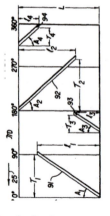 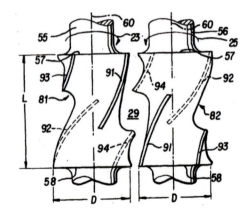

| Unwind rotor geometry | Pair of ST rotors |

Figure 1.38 ST-rotor

These rotors provide new characteristics of dynamic interaction for achieving improved mixing performance compared to F-4-W rotors. They can be driven in mixers with different rotor speeds, however, they reach their best performance in mixers with synchronized rotation of the rotors (even speed). In this case, the two rotors must be aligned in an optimum orientation to each other.

ZZ 2-Rotor

The ZZ 2 rotor geometry was developed by *Wiedmann and Schmidt* [2] in order to reduce the disadvantages of tangential mixers (poor temperature control, moderate distribution of ingredients) compared to intermeshing machines. The ZZ 2 rotors restrict the flight angles of the main vanes to 40°, shortened their length to approx. more than the half of the rotor length and opened a second connecting passage at that side of the rotor where the wing started (Fig. 1.39). Because of the open passages material streams could be divided before and flow together after the wings. These arrangements guide to a better material flow inside the mixer and cause lower heat generation.

The main achievement of this development work was a rotor with a considerably better distributive mixing capability. The rotor shape also significantly improved the control of compound temperature in the mixer. Comparative testing of technical rubber compounds indicated reductions of up to 30 °C in discharge temperatures compared to other 4-wing rotor geometries. Not only can the mixer operate at higher rotor speeds before reaching the same discharge temperature, considerable reductions in mixing time are also possible. Comparative tests with the N-rotor have shown throughput improvements of up to 30 %.

Because of his excellent temperature control and the very good distribution behavior of ingredients inside the batch, the ZZ 2-rotor is typically used in mixers for the

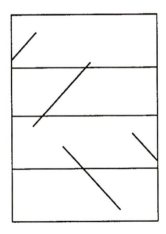 

Figure 1.39 ZZ 2 rotor

technical rubber goods industry, where compounds are more and more produced in one stage, and in final batch mixers in the tire industry.

HDSC-Rotor (High Dispersion Super Cooled)

To improve dispersive mixing, the long wings in the HDSC-rotor where extended to 80 % of the rotor length (Fig. 1.40). One main wing is placed on the drive side of the rotor and the other on the water side. Two little wings ahead of the long wings on the opposite side of the long wings form a passage through which the material has to flow. This wing arrangement supports the flow over the wing tips and improves the dispersion of the compound. Because of their aggressiveness, HDSC-rotors are installed in master-batch mixers of the tire industry.

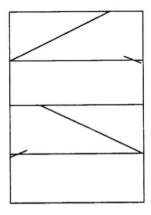 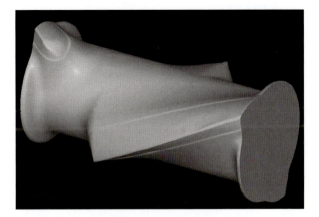

Development drawing   CAD-Model of an unfinished rotor

Figure 1.40 HDSC-rotor

MDSC-Rotor (<u>M</u>aximum <u>D</u>ispersion <u>S</u>uper <u>C</u>ooled) [10]

The MDSC-rotor is equipped with 4 long wings which are uniformly distributed on a core area around the rotor, and 2 wings on both sides of the rotor, respectively (Fig. 1.41). The length and the angle of all 4 wings are the same. A spiral cooling inside the rotor provides good thermal performance. Therefore, the rotor is labelled "Super Cooled".

Practical tests with different tire compounds have shown that the new MDSC-rotor is a very aggressive rotor. A lot of energy can be introduced into the compound and low viscosities of the mix can be reached in a short time. Therefore, this rotor is particularly suitable for master-batch mixing. High productivity can be reached in tire compound production together with good dispersion. Therefore, this rotor is used in master-batch mixers in the tire industry.

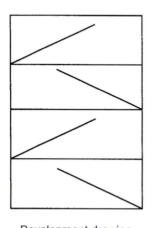

Development drawing   CAD-Model

Figure 1.41 MDSC-rotor

Six-Wing Rotor [11]

The 6-wing rotor is a one-piece cast steel rotor with a single large cored cavity and a conventional cooling system inside. Three long wings extend spirally around the central axis, set apart from each other by an equivalent angle in circumferential direction (Fig. 1.42). Three short wings on the opposite side of the rotor prevent material from remaining at the rotor end plate of the mixing chamber. The clearance between rotor tips and chamber vary over the wing length and from wing to wing.

With its three long wings the 6-wing rotor exhibits good mixing performance. Compared to the standard 4-wing-rotor, Mooney viscosity can be reduced faster and mixing cycles are shorter while providing good compound quality.

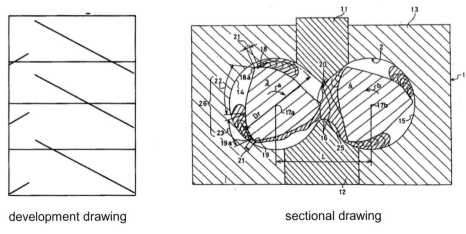

development drawing   sectional drawing

Figure 1.42 6-wing rotor

1.8.4 Rotors for Intermeshing Internal Mixers

History of Development

The age of intermeshing technology in internal mixers for the rubber industry started with the invention of *Cooke* in 1934 [1]. He filed a patent application for an internal mixer with intermeshing counter-rotating rotors (Fig. 1.43). Every rotor has a long helical wing with a compartively wide peripheral face concentric with the axis. This main wing is extending over the full length of the rotor and passing half way around the rotor. Two additional short wings are placed on the opposite half of the rotor on each end. During counter rotation of the rotors, the long wing moves freely between the two short wings of the opposite rotor. The clearances between the two wings are the same as those between the wing tips and the mixing chamber.

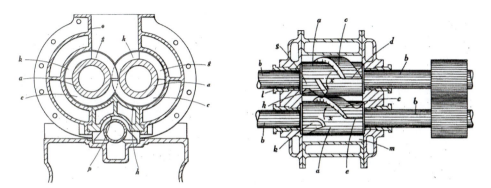

Figure 1.43 First internal mixer with interlocking rotors by Cooke [1]

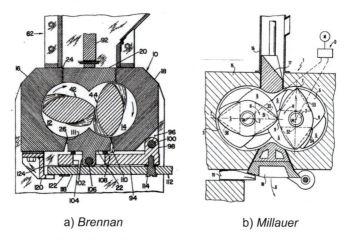

a) *Brennan* b) *Millauer*

Figure 1.44 Internal mixer with interlocking rotors by Brennan [12] and Millauer [13]

Some months later after Cooke's invention, *Lasch and Frei* [7] filed a second patent application for an internal mixer with intermeshing rotors. During the next decades no further patent applications were filed.

Only at the end of the 1960s, *Brennan* [12] invented an intermeshing co-rotating internal mixer (Fig. 1.44), which was derived from intermeshing co-rotating twin screw extruders. *Millauer* followed in the 1970s with his invention of an intermeshing rotor with four wings (Fig. 1.44). In comparison to *Cooke's* rotor, the wing tips of Brennan's and Millauer's rotors are small, like the tangential rotors. Both systems have never been commercially successful.

Interlocking Technology

During the 1980s, *Johnson* and his co-workers [14] developed a rotor based on *Cook's* rotor with two counter-rotating rotors with parallel axes (Fig. 1.45), which was installed in many internal mixers. Each rotor is supported with a long wing of generally helical formation and a wide wing tip that starts at one end of the rotor and builds up a passage on the other end. Two smaller wings are radially spaced before and after the long wing, starting on each end of the rotor. During rotation the wings of each rotor are gearing with the other rotor. In Fig. 1.45, the wings of

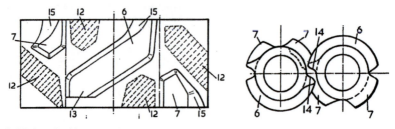

Figure 1.45 Interlocking rotor by Johnson et al. [14]

the parallel rotor can be seen in the manner of footprints beneath the wings (hatched area). Interlocking technology is also characterized by the radial clearance between the core area of one rotor and the tip of the opposite rotor. In this case, it has to be bigger than twice the clearance between wing tip and mixing chamber.

PES-Technology

PES stands for the German name "**P**artielle **E**volventen **S**chaufel". This kind of intermeshing rotor was invented by *Wiedmann and Schmidt* [2] at the end of the 1070s and and underwent further improvements until today.

PES 1 Rotor

The PES 1 rotor resembles the tangential ZZ 2 rotor; both rotor patents were filed in the same patent application (ZZ 2 rotor = first embodiment; PES 1 rotor = Third embodiment). Instead of two long wings on the ZZ 2 rotor, the PES 1 rotor has only one long wing in the axial middle with a passage at each end. Two small wings are radially placed before and after the main wing at the two ends of the rotors.

The PES 1 rotor was not very successful. Only two machines were built (Fig. 1.47) and tried in an internal mixer producing compounds for technical rubber products.

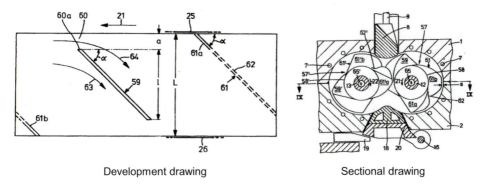

Development drawing　　　　　　　　Sectional drawing

Figure 1.46 Intermeshing rotor by Wiedmann and Schmid [2]

Figure 1.47 PES 1 rotor geometry

PES 3-Rotor

To improve the mixing effect of the PES 1 rotor, the width of the wing tips were increased significantly so that more material will be squeezed between the counter-rotating rotors (Fig. 1.48). This provides better dispersion because the carbon black pearls are destroyed under the high elongation stresses inside the batch between the rotors.

In this machine, a relatively long main wing carries the compound from the center towards the chamber ends. Two short wings are arranged so that the long wing of the opposite rotor intersects with the two short wings. The short wings carry the mixture from the chamber ends back to the centre. If the long main wing were to run right up to the chamber end, dead zones would form in the corners and the compound would not participate completely in the mixing process. Truncations at the ends of the long wings mean that dead zones in this area are avoided. The compound flows past the end of the wings and fully rejoins the mixing process.

Figure 1.48 PES 3-Rotors

PES 5-Rotor

A further increase in mixing efficiency was achieved with an advancement of the PES 3 rotor called PES 5 rotor in the beginning of the 1990s. Here, through numerous laboratory tests and in collaboration with partners in the rubber industry, the PES 3 rotor geometry was modified such that throughput improvements of up to 15% became possible, depending on the compound type. Tests in production mixers with free volumes ranging from 135 to 320 liters have since confirmed these results. This means that it can make very sound economic sense to replace old PES 3 rotors with the new PES 5 rotors.

PES 6-Rotor

Because of the immense cost pressure in the rubber industry, productivity is the main target for improvements of internal mixers. This pressure led to the development of the PES 6 rotor geometry in 2007. This new rotors allow a further increase in productivity of about 10 – 15 %, depending on the kind of compound. However, to be able to use these advantages, adequate installed power is necessary and in some cases mixing procedures have to be adapted.

1.9 Mixer Base Plate

1.9.1 Design

The function of the base plate is to absorb the forces bearing upon the individual components (rotors, end frames, mixing chamber halves) during the mixing process and to deflect them in part into the foundations of the mixer. The base plate should have high torsional rigidity and low self-weight.

The drop door of the mixer and its locking mechanism or latch assembly are incorporated in the base plate. Here too, the modular design is employed (Fig. 1.49). The drop door and the latch assembly are individual modules within the base plate that can be removed and replaced at any time.

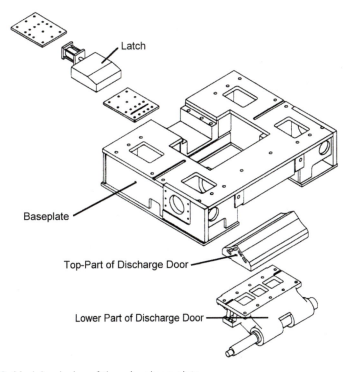

Figure 1.49 Modular design of the mixer base plate

1.9.2 Drop Door and Latch Assembly

The drop door consists of two components, the drop door top and the drop door base. The drop door base is mounted on the mixer base plate on a shaft which in turn is fitted with a hydraulic rotary actuator to open and close the drop door. The drop door top is screwed onto the drop door base and adjusted to fit the mixing chamber during mixer assembly.

Figure 1.50 shows a lateral sectional view through the lower part of an internal mixer. A dotted circular arc shows the path taken by the drop door when opening or closing. The drop door is tightly locked with a latch that is moved in or out by a hydraulic cylinder. Where the latch contacts the underside of the drop door base and applies the locking pressure, a replaceable wear strip is fitted to extend mixer life. To open the drop door, the latch is completely withdrawn and the door opened by the hydraulic rotary actuator. With the drop door open, the contents of the mixer can fall out without any obstruction.

The drop door top, as with all other mixer components coming into contact with the compound, is hard coated for wear resistance and can be temperature controlled. The temperature control media is fed through the drop door shaft via a rotary joint.

The seal of the drop door in the mixing chamber halves and the lateral wear plates is provided by the direct contact of the various components with one another. This is known as a metallic seal and is achieved during the final mixer assembly when the contours of the components are adjusted to make them a perfect fit.

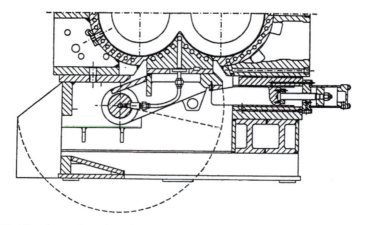

Figure 1.50 Mixer base plate, drop door and latch (sectional drawing)

1.9.3 Drop Door and Toggle

Another possibility to keep the drop door closed is a toggle. The toggle system is mounted on the base plate of the mixer. It is mounted in a hinge and moved by a hydraulic cylinder (Fig. 1.51).

The toggle moves in when the drop door opens (Fig. 1.52 a) and hinges out when drop door is closed (Fig. 1.52 b).

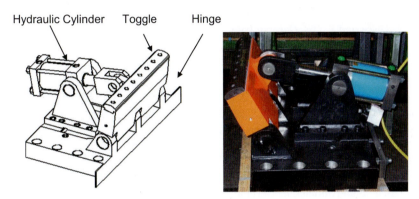

Figure 1.51 Toggle in driven out position

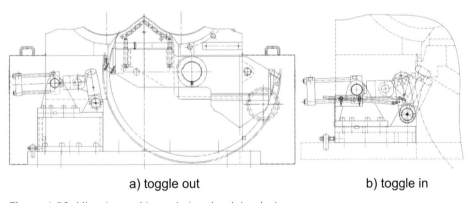

a) toggle out   b) toggle in

Figure 1.52 Mixer base with toggle (sectional drawing)

References

[1] R. T. Cooke: British Patent (filed June 14, 1934) 431,012 (1935)
[2] W. Wiedmann and H. M. Schmidt: German Patent 28 36 940 C2 (1982)
[3] J. Peter et al.: Tandem-Mischverfahren; KGK 47 (1994), P. 656
[4] F. H. Banbury: US Patent 1,200,070 (1916)
[5] F. H. Banbury: US Patent 1,227,522 (1917)
[6] F. H. Banbury: US Patent 1,818,449 (1931)
[7] A. Lasch and K. Frei: German Patent 738,787 (1943)
[8] D. Z. Tyson and L. F. Comper: US Patent 3,230,581 (1966)
[9] N. O. Nortey: US Patent 4,744,668 (1988)
[10] A. Limper et al.: European Patent 1,649,995 (2008)
[11] K. Takakura et al.: European Patent 0,774,331 (2003)
[12] A. K. Brennan: US Patent 3,490,750 (1970)
[13] Ch. Millauer: US Patent 4,084,263 (1978)
[14] F. Johnson et al.: European Patent 0,170,397 B1 (1990)
[15] Lasch, A. and Stromer, E.: German Patent 641 685 (1934)

2 Processing Aspects of Rubber Mixing

A. Limper

The mixing of rubber compounds is a sophisticated task. Many components are difficult to dose and materials are delivered to the mixer in all thinkable forms, as rubber bales, oils, powders, hard resins, granules, chips, and even pastes. As the conversion of these raw materials into a free flowing form is quite expensive, the discontinuous mixing by a kneader (or batch mixer) is still the most versatile and most economic solution.

To prepare a perfect batch, different kinds of mixing are applied:

- distributive mixing
- dispersive mixing
- laminar mixing

■ 2.1 Mixing Principles

Distributive mixing is shown in Fig. 2.1. Here, the particles at the beginning of the mixing process are not distributed in the cross section. Distributive mixing means that particle positions are changed to reach a uniform distribution in the batch. Note that particle size does not change during this mixing procedure.

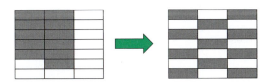

- No Change of Particle-/Domane-Size
- Position Changes

Figure 2.1 Distributive mixing

Driving forces for position changes are consequently the rotations of the material within the mixing chamber, the intensity of the material transport phenomena in the mixer, and the total number of rotations applied to the batch. It is also clear that all parts of the dosed ingredients have to be part of the mixing process so that no dead spots occur in the mixer.

Dispersive mixing is shown in Fig. 2.2. Other than in distributive mixing, here a change of the particle size also occurs. The size reduction is normally achieved by the application of shear or strain forces. Practical studies show that in order to achieve particle size reduction, straining is much more effective than shearing. As in a normal batch mixer, a lot of strain effects take place (e.g., passage through gaps between rotor blade tip and wall or through the gap between the rotors), which makes for a good dispersive mixer. Driving forces for this kind of mixing are applied forces and the application time. They can be roughly characterized by the applied torsional moment or the power demand of the mixer.

Figure 2.3 shows the characterization of particle size reduction as a function of applied forces and time. As it can be seen, particles can resist a certain (lower) stress for an indefinite time. In other words, if the stress level in the mixer is low (e.g., if due to high material temperature the compound viscosities are low), longer mixing times do not help to achieve good dispersive mixing.

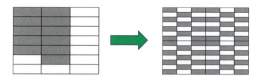

- Particle-/Domane-Size Object of Major Change
- Position Changes Might also Occur
- Shear Stress, Time

Figure 2.2 Dispersive mixing

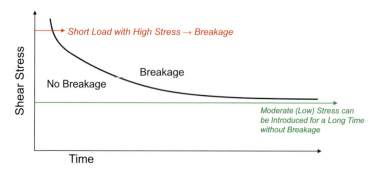

Figure 2.3 Particle size reduction as a function of applied force and time

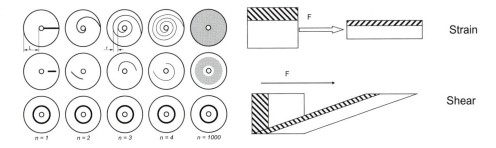

- Shear Stress τ
- Shear Angle γ; Shear Rate γ̇

Figure 2.4 Laminar mixing

At higher stress levels an immediate breakage can be realized. It means that for good dispersive mixing a short impact of high stresses is very effective. In practical terms, a high power peak at the beginning of a masterbatch mixing cycle can be used to achieve an immediate high amount of dispersive effects. The figure also shows a correlation between application time and breakage.

The third mixing mechanism – *laminar mixing* – is shown in Fig. 2.4. Here, the interface between two layers should be increased as much as possible. The increase can be achieved either by strain or by shear. It also can be seen that rotational shear has a high effect. Laminar mixing is of particularly high importance for rubber. During dwell time in the mixer, material layers can be stretched by several hundred percent and repetitive straining and stretching provide high laminar mixing effects. The total shear/strain ratio is used to characterize laminar mixing. It can be calculated, or at least estimated, by the respective integral of the shear/strain rate as a function of time.

The interrelation of all three kinds of mixing with the practical process parameters are shown in Fig. 2.5. The dispersive mixing is a function of position changes, which can be characterized by the rotor speed and the applied time. Here, only the times when the ram is acting on the compound should be considered.

The breakage of particles during dispersive mixing is achieved by the applied forces, which can be characterized by torque (or the power at a given speed) and the application time. These forces can be determined by the "fingerprint" of a mixing cycle, which typically shows the power as a function of the mixing time.

Laminar mixing is characterized by the absolute deformations introduced into the batch. They are characterized by the passages through the gaps between the rotor flights and the mixer walls and the passages through the rotor gap area.

1. Distributive Mixing (Position Changes)
   → Revolutions   (Rotor Speed, Time)

2. Dispersive Mixing (Breakage by Shear Forces)
   → τ_{max}; Time t; τ_{mean} (Power, Torque)

3. Laminar Mixing (Inter-Surface Increase by Deformation)
   → Passages through Gaps (Rotor/Wall; Rotor/Rotor)
                                   (Ram Pressure, Revolutions, Fill Factor)

Figure 2.5 Type of mixing and process parameters

■ 2.2 Process Description

Figure 2.6 shows a rotating rotor in a batch mixer [1, 2]. The material to be mixed is positioned in front of the red side of the flight and can follow three flow paths:

- it can be transported rotationally with the rotor flight ("rolling bank")
- it can overpass the gap between the wing tip and the chamber wall
- it can be pushed along the flight and pass around the rotor end

The "rolling bank" is a highly active mixing zone. Similar to the mixing process on roll mills, here material layers are opened and folded and a lot of energy is introduced into the material. Due to the flow paths mentioned (paths 2 and 3), the size of the rolling bank decreases during a rotor revolution.

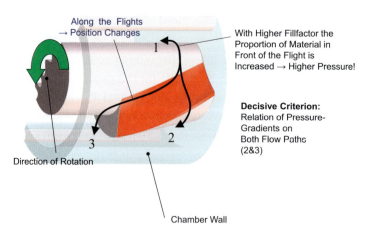

Figure 2.6 Flowpaths within the mixer

The decisive criteria for the partitioning of the flow between the paths along and over the rotor wings are the pressure losses along both. If the mixer is poorly filled, a passage along the flights is easy and thus the mixer is pushing the material along its flight to edges of the mixing chambers. As a consequence, the transport of the "rolling bank" ends after a short angle of revolution. Therefore, in this process phase almost no material is transported from the bottom part upwards by the fights. It means that the ingredients more or less stay at the bottom part of the mixing chamber. This is often seen in the first steps of a mixing cycle, when the mixer is still underfilled. In this phase the mixing process is primarily distributive while dispersive mixing is poor.

With increasing filling degree, flow paths along the flights become more and more blocked and the material begins to pass over the gap between the flight tip and the chamber wall. Here, high shear stresses are exerted so that with an increasing fill factor the mixer changes its characteristics from distributive to dispersive mixing, see Fig. 2.7.

Figure 2.8 summarizes all described effects in terms of general process parameters [10 – 14]. The two diagrams summarize systematic trials on a tangential internal mixer, on which an SBR/N 220 compound (typical tire material) was mixed. The y-axis indicates the dispersion quality (note that smaller values mean a higher quality!), while the x-axis shows fill factor of the mixer. This value stands for the percentage to which the volume of the machine is filled by the compound. A fill factor of 0.5, for example, stands for a 50 % filled machine.

The left diagram in Fig. 2.8 represents a throughput of 68 kg/h. In all performed trials mixing time and fill factor were adjusted to reach this value. Every line stands for a specific rotor speed applied in the trials: 25, 38, and 63 revolutions per minute, respectively. The results show that with an increasing fill factor the degree

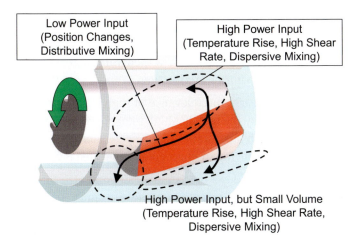

Figure 2.7 Transition from distributive to dispersive mixing

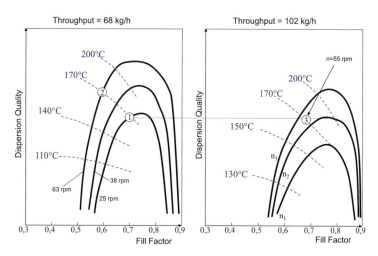

Figure 2.8 Optimization of throughput / dispersion quality

of dispersion increases, because of the conversion of the mixing procedure from pure distributive mixing at low fill factors to more and more dispersive mixing at higher degrees of filling of the mixer, as mentioned earlier.

With increasing fill factors, the appearance of "dead spots" in the mixer prevents better dispersion results. As a consequence, it can be assumed that there is an optimum fill factor for every mixing cycle with regard to dispersion quality. The interrupted lines also indicate the drop temperatures of the batches. It can be shown that with both increasing rotor speed and increasing fill factor the temperature rises.

These general relationship allows for a systematic improvement of the mixing cycle. If the starting point is process no 1 in the left diagram, an optimization can be achieved by a simultaneous decrease of the fill factor and an increase in rotor speed (process no 2). Thus a significant improvement of the quality could be achieved at constant throughput and constant drop temperature.

Alternatively, beginning with process 3 means a 50% higher throughput at a constant drop temperature is achieved by a moderate increase of rotor speed, a decrease in fill factor, and a shorter mixing time.

The diagram also shows clearly the relevance of high stress levels for the dispersion quality, demonstrated by the fact that high rotor speeds lead to much improved quality levels. However, with a constant mixing speed along the total mixing cycle the drop temperatures are moving to unacceptable elevated levels.

Using variable speed drives allows to systematically optimize the mixing cycle. It means to start with high rotor speeds to achieve a quick break down of filler particles and then to move to moderate speeds for a slow temperature development. The latter result in mixing times that allow a homogenous dispersion quality within the batch (which cannot be expected at extremely short mixing times).

Both diagrams in Fig. 2.8 show in an impressive way how the global parameters of mixing can be systematically improved. The example of variation of the rotor speed also shows that parameter variations along the mixing cycle allow an even more intensive optimization of the process. To achieve this, operators need more information on the status of the actual process. One measure for a systematic approach of such an improvement is the use of process data which are recorded during the mixing cycle. One example is the so called "fingerprint", which normally shows the power demand of the mixer as a function of mixing time.

Figure 2.9 shows a typical "fingerprint" of the mixing cycle for a masterbatch compound [1 – 4]. First, the polymer (in this case natural rubber) is dosed into the machine. After a first and short power peak, the power demand of the mixer is reduced due to the mastication of the rubber (resulting in a viscosity reduction) and due to the temperature increase. After about 30 seconds of pre-mastication, the ram is lifted (resulting in a short power minimum) and carbon black is added to the mixer. When the ram is lowered again, power demand exhibits a sharp increase. This is related to the higher fill factor of the machine.

Addition of high filler volumes can lead to a substantial temperature decrease of the compound. After the addition of the fillers and the lowering of the ram, many contradicting effects can be observed.

We generally see

- an increase of temperature → resulting in lower viscosities
- an increase of the fill factor of the mixer, as the ram is only slowly lowering itself → this alone leads to higher power demands
- an increase of viscosity of the compound as it is transferred from a two-phase system ("rubber + carbon black") into a monophase "rubber compound". Incorporation of carbon black causes an increase in viscosity and thus an increase in power demand.

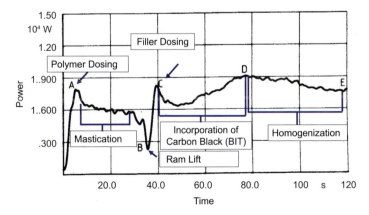

Figure 2.9 Fingerprint of a typical mixing cycle

In summary, the power demand shows a maximum (point "D" in Fig. 2.9). The time span between the power peaks "C" and "D" is often called "Black Incorporation Time" (BIT) [5 – 9], but as mentioned above, it summarizes many effects – not only the incorporation. It also does not apply to all kind of recipes. However, it is a good measure to compare similar mixing procedures as long as the mixing processes take place under similar process conditions (rotor speed, order of dosing, etc.).

Although the pure "power-print" provides only limited information, it is quite useful to also look at other process parameters. In the last years, mixers have been equipped with more and more sensors. A very important parameter to judge the mixing process is the ram position as a function of time. Figure 2.10 shows the fingerprint of both parameters for a mixing cycle. It can be seen that after the polymer addition, the ram immediately reaches its end position. Immediately after the dosing of the carbon black, the ram is lowered fast. Then the rotors transport material to the area below the ram and consequently the pressure is increasing. In the shown example this pressure increase it is even able to raise the ram back into the chute. After the ram has passed a maximum position, it slowly approaches its lowest position. In the demonstrated example this is in the vicinity of the power maximum "D".

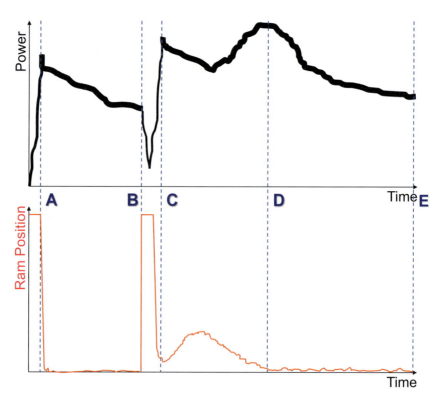

Figure 2.10 Typical process parameters during the mixing of polymer and carbon black

The ram seating time after the addition of fillers can be used as a tool to judge the fill factor of the machine. If the ram comes down too fast, the machine is under filled. A proper time for ram seating (only as a "rule of thumb") is about 30–45 seconds, which should only be higher in highly filled compounds. If the ram does not reach its end position in the mixing time, undispersed filler particles will be dropped. It also should be noted that the time when the ram reaches its lowest position is the beginning of the mixing process for some particles. It means the operator should allow for sufficient mixing time after the ram reaches its lowest position to make sure these particles are dispersed. The same is also valid for ram cleaning steps. Ram cleaning means that considerable amounts of fillers are brushed into the mixing chamber. They must also be incorporated properly. Therefore, after a ram lift or after the ram is finding its final position, as in the example shown, a mixing time of 45–60 seconds is typically needed.

Finding an optimal fill factor should start with a slightly "overfilled" machine. By measuring the ram position at the end of the mixing cycle, the volume remaining "unmixed" below the ram can be calculated and subtracted from the next batch. Thus, a good fill factor can typically be found quickly.

Besides the ram seating time (see above), the fill factor can be judged by the movement of the ram in the last seconds of the mixing cycle. If the mixer is properly filled, the ram should "dance" around its final position. This effect can be seen very clearly for intermeshing mixers and is also quite pronounced for tangential rotors with long wings (e.g., full-4 wing; H-swirl; ST etc). It might be less visible for "soft acting" tangential rotors, e.g., ZZ2-types.

■ 2.3 Influence of Raw Material Properties

Although the process parameters have a strong influence, also the raw materials play a major role in the definition of the compound properties [16–20]. Figure 2.11 shows the influence of the temperature on the viscoelastic properties of natural rubber [1]. It can be seen that the torque of the vulcameter (which is characteristic for the viscosity of the rubber) shows an exponential decrease with increasing temperature. A comparison, e.g., of the viscosity values for 30 and 50 °C shows a drop of 30%.

In a typical mixing room, even bigger temperature differences in the polymer fed to the mixer will occur. Here, temperatures of below 0 °C in the wintertime and 50 °C at the end of a hot summer are possible. If the polymer is fed to the mixer without any temperature regulation, the power demand of the mixer can vary significantly (such as, e.g., during carbon black dispersion). To achieve consistent quality, the processor should keep the dosing temperatures as constant as possible.

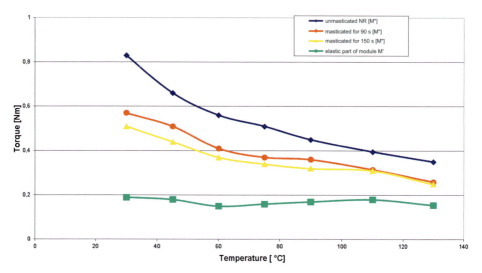

Figure 2.11 Change of viscoelastic properties as a function of mastication time for natural rubber

In an ideal case, this is guaranteed by temperature controlled storage, in which the pallets of polymer are stored for a certain minimum time (e.g., 100 hours).

The consequences of inconsistent feeding temperatures are shown in Fig. 2.12 [1]. Here, the polymer was fed to a lab mixer and the maximum power demand in the mastication phase was measured. As the picture shows, an increase of feeding temperature of the polymer from 20 to 60 °C leads to an approximately 40% lower power peak.

The average power demand in the mastication phase is also dependent on the size of the polymer lumps dosed to the mixer. As Fig. 2.13 shows, the energy input in the mastication phase can vary significantly. In the shown example, polymer pieces of different sizes were dosed to a lab mixer. The small particles lead to a low energy input, because they are able to "escape" the rotor flights and thus prevent an effective power input. When parts are too big, they are only dragged in with a low velocity and with a fixed mastication time, the power input is low. Figure 2.13 also shows that for medium sized parts the power input is at a maximum level.

In the practical use of internal mixers this effect is also visible. When polymer bales are cut for a proper weight, all the small cut-off pieces are fed to the mixer from time to time. For sensitive compounds the different power demand, which follows this inconsistent feeding, can lead to batches out of specification.

These examples of the parameters polymer temperature, -form, and -size show the influences of raw material properties on the result of mixing. Other raw material parameters also play an important role and a discussion will follow in this book, see Chapter 5.

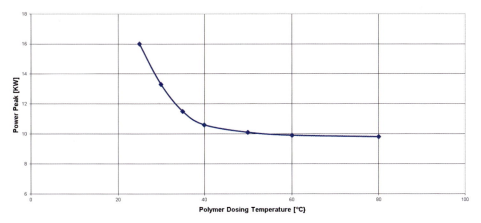

Figure 2.12 Power demand in the mastication phase for different polymer temperatures

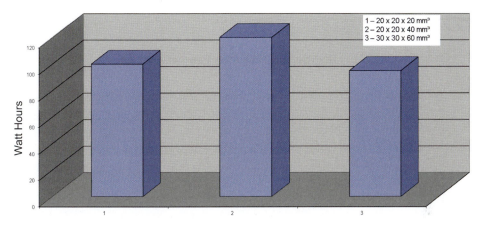

Figure 2.13 Energy demand for different polymer lump sizes

■ 2.4 Influences of Process Parameters

The thermal boundaries of the mixer can change the mixing process significantly. As Fig. 2.14 shows, the drag-in process of polymers is heavily dependent on the forces on the surface of the rotors. The surface conditions (slip < > stick) are very much influenced by the contact temperature, which in turn is influenced by the rotor temperature.

Figure 2.15 shows that the ram seating time after polymer feeding can be heavily influenced. In this example, a lab mixer was fed with polymer and the time to bring the ram to its lowest position ("ram seating time") was measured. The maximum time was limited to two minutes. As the figure shows, the ram seating time

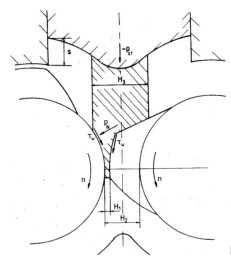

Figure 2.14 Drag in process in the gap region

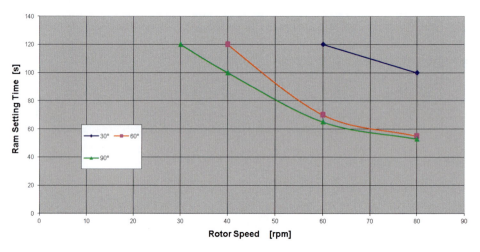

Figure 2.15 Change of drag-in conditions by different TCU settings

decreases with increasing rotor speed. Also, an increasing mixer temperature leads to significantly lower ingestion times. This effect is most pronounced for a change from 30 to 60 °C. As the wall sticking is almost complete at 60 °C, a further increase in temperature (up to 90 °C) only leads to a slight change in drag-in conditions.

During the phase of carbon black incorporation and dispersion, the temperature level of the mixer also plays a decisive role. As Fig. 2.16 shows, the BIT of an

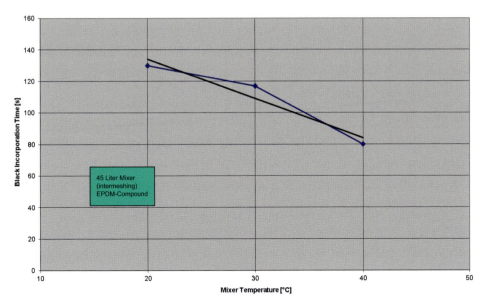

Figure 2.16 Change of BIT by different TCU settings

EPDM/N550 blend decreases sharply, when the temperature of the machine is increased from 20 to 40 °C.

The examples show that the thermal boundaries have a substantial influence on the process conditions. They also explain the existence of "first batch effects" in practice. Here, at the start-up of a mixer, the first batches might show other qualities than the following batches, because the mixer starts to work at rather low temperatures and is heated up to quasi-stationary wall temperatures. This means, under the pre-condition of constant mixing times, the start temperatures for every batch are the same. The time to reach such a quasi-stationary equilibrium can take up to 15 batches. Solutions to eliminate this effect include:

1. heating up the mixer walls to the "quasi-stationary" temperature level before the first batch and to switch to "normal cooling" just after the start of mixing.
2. using some cleaning batches to heat up the mixer quickly.

Both methods should also be applied after disruptions in the production process. To avoid cooling-down the mixer too quickly, at least the cooling should be switched off at longer interruptions.

Figure 2.17 shows – in general terms – the influence of the thermal boundaries on the dispersion quality and the drop temperature [14]. Here, a significant change in wall and rotor temperatures (30 °C) only leads to inferior minor change in drop-out temperature. The lines also show the increase of quality with increasing machine temperature (which was explained earlier). It can be summarized that the influence of the wall temperature on the drop temperature is visible, but it is not as

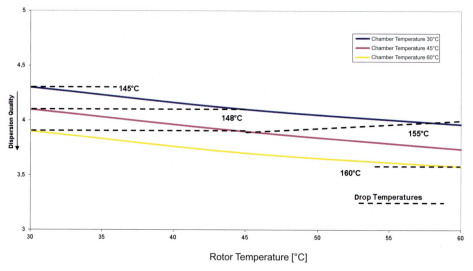

Figure 2.17 Dispersion quality for different thermal boundary conditions

pronounced as expected. Processors should rather use the speed of the rotors to change temperature levels during mixing. Wall temperatures should be adjusted as high as acceptable to allow quicker drag-in processes, faster carbon black incorporation, and the highest possible power input during the first phase of mixing polymer and fillers.

Figure 2.18 points out that the ram pressure is also a decisive process parameter [14]. Here, the dispersion quality and the optimum fill factor are increasing with increasing ram pressure. As the drop temperature rises with the increasing ram pressure, the processing window becomes narrower.

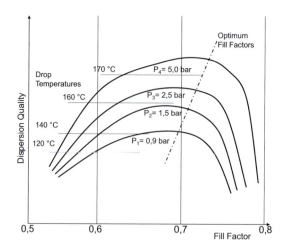

Figure 2.18 Influence of the Ram Pressure on the Mixing Process

The results also show that non-constant pressures will always lead to inconsistent results of the mixing process. This fact is one reason for the introduction of hydraulic rams into practical use. Pneumatic rams, which had been the standard for the ram action for decades, work with unsteady pressure levels. When a lot of pressurized air is consumed (e.g., in the day shift with a high demand of the compressed air supply) either the pressure level itself is lowered or the time span for a recovery of the pressure after a ram stroke is significantly elongated. Hydraulic rams have their own station for each mixing line and thus provide more constant conditions.

Figures 2.19 to 2.21 show power curves (fingerprints) for hydraulic rams for different fill levels. First, Fig. 2.19 shows a standard mixing process for a compound for technical rubber goods.

Within approx. 150 seconds a (single-step) compound was produced with a fill factor of 76% (in a 45 L intermeshing machine). The ram position curve shows a good seating characteristic, as it takes about 45 seconds to reach the final position after all ingredients have been added.

Figure 2.20 shows the "same" process, but with a 10% increased batch weight (which means 84% fill factor). The curves show that the ram is not able to reach its end position, even after about 70 seconds it is still some mm above the total ram down position.

After the ram pressure was raised from 60 N/mm² (Figs. 2.19 and 2.20) to 100 N/mm², the ram action shows a proper behavior: The ram is again down in about 45 seconds (Fig. 2.21). To limit the temperature increase, in this case the rotor speed was lowered in the very last phase of the mixing cycle. In both cases (Figs. 2.19 and 2.21), the quality reached the same level. The example shows, that

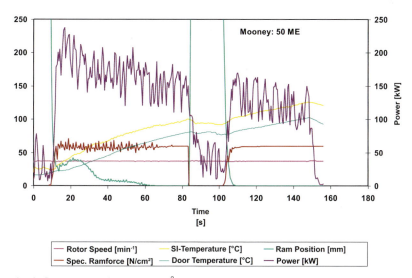

Figure 2.19 Power curve for 60 N/cm² ram pressure and 76% fill factor

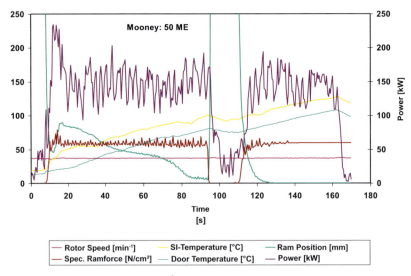

Figure 2.20 Power curve for 60 N/cm² ram pressure and 84% fill factor

with an intelligent use of the possibilities of the ram, the output can be increased by 10%.

Rising ram pressures, however, also poses some risks which have to be noted. When the ram is pushed onto the dosed material, high ram pressures increase the flow of the polymer (or of the compound) into the gap between the ram and the feeding chute. If this area is filled, the ram movement is not only influenced by the dragging-in process of the rotors, but also dependent on the shear forces between the feeding chute and ram.

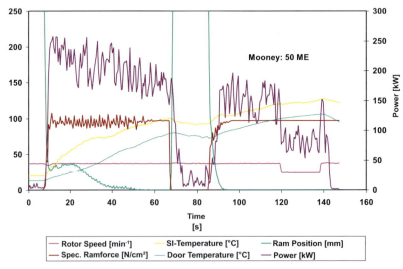

Figure 2.21 Power curve for 100 N/cm² ram pressure

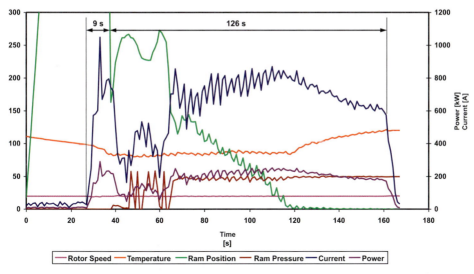

Figure 2.22 Example of a Ram Jamming (Second Mixing Stage) for a 135 L Intermeshing Mixer

These forces can be so high, that the ram is not able to move anymore ("ram jamming"). An example for this is shown in Fig. 2.22, which depicts a final mixing process. First the masterbatch is fed into the 135 L intermeshing machine. About 10 seconds after the masterbatch is completely dosed, the ram was moved down. The very long time span needed to bring the ram to its final position (about 80 seconds!) is a hint for "jamming problems" (which the operator indeed faced).

Figure 2.23 shows a process modification that solved the problem. After the dosing of the masterbatch, the ram was sent down with a time delay of about 26 seconds.

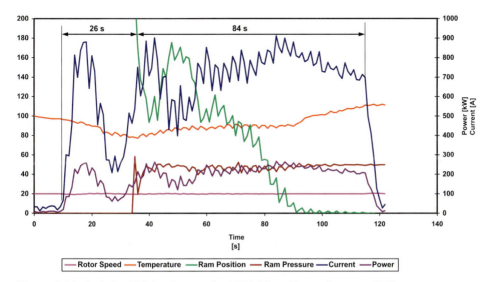

Figure 2.23 Optimized Mixing Process for 135 L Mixer (Output Increase 25%)

Now the ram reaches its end position in about 50 seconds. As the contact time between the masterbatch and the ram was cut by 35%, ram sticking problems could be completely avoided. The saved time for the complete ram movement can be subtracted from the total mixing time and the mixing cycle is approx. 30 seconds shorter (about 25% more throughput).

As mentioned initially, the mixing sequence represents an influence parameter that should not to be underestimated. Figure 2.24 explains using the example of oil addition at different times [12].

In the shown mixing cycle, first the polymer is fed into the mixer. After a mastication time of about 1 minute the oil and carbon black components are added. If – as displayed in the upper part of the figure – the oil and the carbon black are added at the same time, the oil is incorporated quickly, but the maximum torque (needed to achieve maximum carbon black dispersion) is relatively small.

With a delayed oil addition (middle picture) a better dispersion quality can be obtained. However, it is absolutely necessary to choose the time for the oil injection so that enough carbon black surfaces remain available for oil absorption. If this is

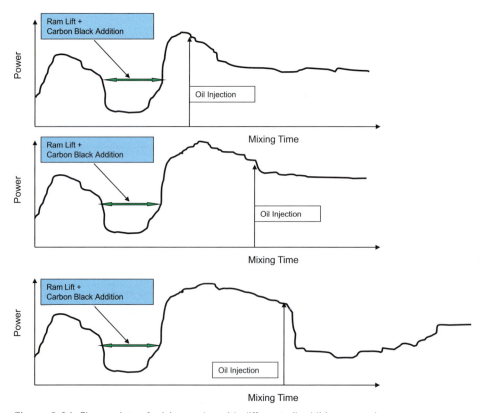

Figure 2.24 Fingerprints of mixing cycles with different oil addition procedures

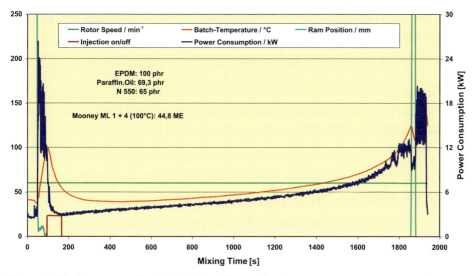

Figure 2.25 Power curve for the injection of large oil volumes

not the case, oil incorporation will be delayed. Lubrication films are created in the mixer and it is hardly possible to bring energy into the compound. This can lead to a considerable increase in mixing time, as shown in the lower part of Fig. 2.24.

Although it is difficult for a machine operator to keep all timings for the addition of carbon black and oil, and for changes from high to low speeds, it was clearly shown in Fig. 2.23 that missing these crucial timings can lead to considerable quality loss with manual operation. Figure 2.25 shows an example where a high amount of oil was dosed to the mixer in one injection. The "drowning" of the mixing process can

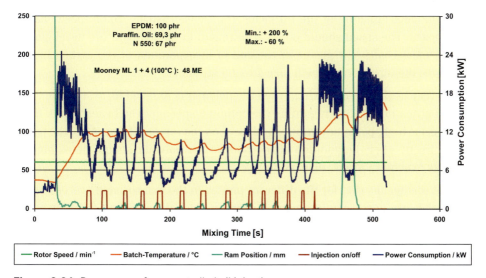

Figure 2.26 Power curve for a controlled oil injection

easily be seen. The mixer needs about half an hour to recover from this problem and the total mixing time is in the range of 1900 seconds [15, 21].

If the oil-dosing is automated in the sense that the injection is stopped when the (power) of the machine has decreased to a critical level and activated only when certain minimal values for the torque are exceeded, the process shows a much better characteristic (see Fig. 2.26). The process is reproducible (which means also constant quality) and the total mixing time is reduced to 520 seconds [15, 21].

■ 2.5 Basic Considerations for the Development of a Mixing Cycle

As the *fill factor* has a decisive influence on the mixing quality, it should be determined first. To do this a complete batch (all its ingredients) is fed to the mixer. The ram movement should be monitored.

- Ram seating times should not be too long
- The ram should be "dancing" in its end position at the end of the cycle. This phase should be at least 30 seconds long.

If the ram does not reach its end position at all (or at a very late point in time), the batch weight must be reduced. To calculate the respective value, the cross sectional area of the chute can be multiplied by the "ram setback position" (i.e., difference between final position and lowest position in the process). Thus, the volume to be subtracted can easily be calculated.

After the determination of the fill factor, the "masterbatch – rotor speed" has to be adjusted. The speed should be set with regard to the temperature development of the "masterbatch" (it means at the end of incorporation of the carbon black). Typically, the incorporation will take 2 to 2.5 minutes.

At the very beginning of mixing, when the polymer is dragged in, systematic trials should give an indication regarding the optimum speed for the fastest input. If the speed is too high, polymer lumps can "dance" on the rotor surfaces. Also, the power peak directly after the addition of the filler to the polymer should be as high as possible to achieve a good dispersion level.

In total this means that for the different phases in the masterbatch process different speeds are necessary (dragging-in; carbon black dispersion; time after power peak).

After consideration of the masterbatch phase, the "final mixing" should be analyzed. Here, the rotor speed is optimum, when an equilibrium of the power input and the heat flow via the surfaces of the mixer can be found. This means that mixing times can be increased without any temperature problem. Often, a compromise has to be found as "equilibrium-speeds" are too low.

If in the final phase, low speeds have to be chosen; also, the total revolutions (to be counted in the phase when the ram is down) should be monitored.

Final mixing – which is normally more a distributive than a dispersive task – needs about 45 to 60 seconds to distribute the chemicals correctly. The downstream equipment also has to be taken into account. If one or even more mills are following the mixer, these aggregates can also introduce some mixing energy.

Some general issues should also be investigated:

Using a "conventional" mixing cycle (it means first the dosing of polymer(s) and then the feeding of fillers), the advantage of pre-mastication of the polymer can be achieved. This procedure is mandatory before adding fillers to natural rubber.

When different polymers are used, the problem of "phase-transition" must be considered. The fillers (especially carbon black) have a preference for a specific polymer to be incorporated in, while the other polymers show lower filler concentrations. In such a case, a "pre-mastication" of the polymers is very useful. Fillers should only be fed after the polymers have formed an interphase system with only small domains of the different partners.

In both described examples, conventional mixing cycles have advantages. The disadvantage of this kind of mixing are:

- When the fillers are added, the polymer(s) has (have) already a certain elevated temperature.
- When the ram travels down after filler addition, compressed air might blow a part of the filler on top of it. Sometimes this necessitates an additional ram lift (cleaning step).
- Fillers, which enter the mixer late (e.g., after ram cleaning) might not be dispersed correctly
- The cycle needs at least two feeding steps
- Chute contamination might be the consequence of the described ram problems.

An alternative to the "conventional way" of mixing is the *"up-side-down"* procedure.

Here, first the fillers are fed to the mixer, then the polymer is added and the ram is driven down. This procedure has the following advantages:

- High power input right at the beginning as the mixer is highly filled and the polymer temperature is low
- The chute remains cleaner than in the conventional process because the polymer seals the ram surface and avoids "blow-up" of the fillers
- The mixing cycle can run with one step less than the conventional mixing cycle.

Disadvantages are of course that the described "pre-mastication" of polymers is not possible. Some fillers also have the tendency to agglomerate in the mixing chamber. In conventional cycles – when the mixer works with polymer chunks that wipe all corners of the mixer – agglomerates can be avoided, whereas during the up-side down process it is hard to break up existing areas of compressed fillers.

At the very end of the process set-up, a "fine tuning" should take place. It means for example:

- Optimization of oil injection (raised speed, time of injection, multi-stage injection, etc.)
- Integration of ram lifts (ram cleaning, avoidance of "dead spots" under the ram etc). Please note that ram lifts should not be used, when a power peak is expected. Also, the residual time after ram cleaning must be as long as necessary to also disperse fillers/chemicals entering the mixer after ram cleaning.
- Speed variations (quick dragging-in or dropping out etc.)
- Temperature settings (e. g., faster dragging-in; -mastication time; -BIT time). The temperature setting also has an influence on eventual problems of residual material in the mixing chamber or sticking material at the drop door. A general evaluation of possible sticking problems can be made when the ram temperature is changed and its lower surface is analyzed.

■ References

[1] F. Grajewski: Untersuchungen zum thermorheologischen Verhalten von diskontinuierlichen Innenmischern zur Kautschukverarbeitung. Dissertation RWTH Aachen 1988.

[2] A. Limper, P. Barth, F. Grajewski: Technologie der Kautschukverarbeitung. Carl Hanser Verlag, München/Wien 1989.

[3] A. Limper, J. Pohl: Gummi-Mischtechnik, Altbewährte Prinzipien und neue Technologien. *Gummi Fasern Kunststoffe*, GAK 9/1992 Nr. 45.

[4] G. Menges, F. Grajewski: Process Analysis of a Laboratory Internal Mixer. *Kautschuk Gummi Kunststoffe* **40** Nr. 5 1987.

[5] N. Nakajima: Energy Measurements of efficient Mixing. *Rubber Chemistry and Technology* Vol. **55** 1982, p. 931.

[6] E. Dizon: The Processing of Filler reinforced Rubber. *Rubber Chemistry and Technology* Vol. **50** 1977, p. 765.

[7] G. Cotton: Mixing of Carbon Black with Rubber III – Analysis of the Mixing Torque Curve. *Kautschuk Gummi Kunststoffe* **38** Nr. 8 1985, p. 118.

[8] G. Cotton: Mixing of Carbon Black with Rubber II: Mechanism of Carbon Black Inkorporation. *Rubber Chemistry and Technology* Vol. **57** 1987.

[9] B. Buskirk: Practical Parameters for Mixing: *Rubber Chemistry and Technology* Vol. **48** 1975.

[10] W. M. Wiedmann, H.-M. Schmid: Optimierung tangierender und ineinandergreifender Rotorgeometrien von Gummiknetern. *Kautschuk Gummi Kunststoffe,* **34** Nr. 6 1984.

[11] H.-M. Schmid. Qualitäts- und Produktivitätssteigerungen im Innenmischer. Vortrag VDI Seminar "Der Mischbetrieb in der Gummiindustrie", VDI Verlag, Düsseldorf 1984.

[12] H.-M. Schmid: Produktivitätssteigerung durch Modernisierung von Kneterlinien. Vortrag DKT 1988.

[13] H.-M. Schmid: Maschinen für das Aufbereiten und Verarbeiten von Qualitätsmischungen für technische Gummiwaren. *Kunststoffe* **72**, 1982 Nr. 7.

[14] H.-M. Schmid: Voraussetzungen und Optimierungsmöglichkeiten bei der Herstellung von Kautschukmischungen im Innenmischer. Dissertation RWTH Aachen 1992.

[15] A. Limper, R. Grünheck: Kosteneinsparung im Mischsaal. In: Kostensenkungspotenziale in der Gummimisch-Industrie. VDI Verlag, Düsseldorf 1993.

[16] J. Hopf: "Einfluß der Eingangsqualität – Wie soll eine gute Rohstoffeingangskontrolle aufgebaut sein?" in "Mischungsherstellung – der Kernbereich der Elastomerverarbeitung"; VDI-Gesellschaft Kunststofftechnik. – Düsseldorf: VDI-Verlag, 1998.

[17] A. Limper, H. Keuter: The Influence of Raw Material Parameter Variations to the Mixing Process and Product Properties in: "A Review of European Rubber Research in Practice – Mini Derucom and Prodesc"; Conference proceedings of the international conference, held in Paderborn, Germany, on January 9th and 10th, 2002.

[18] A. Limper; H. Keuter: The Influence of Raw Material Parameter Variations in the Mixing Room: Long Chain Branching of EPDM Polymers in: "A Review of European Rubber Research in Practice – Mini Derucom and Prodesc"; Conference proceedings of the international conference, held in Paderborn, Germany, on January 9th and 10th, 2002.

[19] A. Limper, H. Keuter, C. Rüter: The Influence of Raw Material Parameter Variations in the Mixing Room: Carbon Black Fines Content – Part I; *KGK* **5** (2003), p. 250–257.

[20] A. Limper, H. Keuter: Quality Assurance in the Rubber Mixing Room. Institut für Kunststofftechnik, Universität Paderborn, Gupta Verlag 2003.

[21] A. Limper, W. Häder: Der Mischsaal unter Systemaspekten: *Kautschuk Gummi und Kunststoffe* 1992 Nr. **9**.

3 Mixing Characteristics of Polymers in an Internal Mixer

M. Rinker, A. Limper

In this chapter, the authors try to describe the different principles governing the mixing process. It should be emphasized that in many cases there are also specific requirements and prerequisites for the polymer itself to provide proper mixing results. This chapter shows some examples for the most processed types of polymers used in the rubber industry.

■ 3.1 Natural Rubber (NR)

Natural rubber still represents about one third of the polymers used in the rubber industry. In its very basic form it has some very high molecular weight portions and consequently very high viscosity. In this condition it cannot be processed or even mixed. Therefore, it first has to be brought to a certain viscosity level by so called "mastication".

Mastication more or less means a reduction of the chain length of the polymer. The processor can decide whether to buy already masticated rubber at a defined viscosity or to masticate the raw polymer in house. Quite often RSS1, RSS3, STR20 or SMR 20 rubber types are used. Viscosity tests have shown that over a period of one year the processors have to cope with a wide range of viscosity fluctuations of more than 20 Mooney Units.

The mastication can be driven by high shear forces, thermal degradation or it can be supported by the help of special chemicals (e.g., peptizers). Degradation at high temperatures (above 160 °C) normally inhibits negative effects on the final products properties. Two other properties also have a big influence on the mixing process: one is the effect of crystallization and the other is the high dependency of polymer viscosity on temperature.

As NR is able to crystallize, it shows that a partial crystallization can be initiated after storing in cold areas. For example, this might happen below 15 °C, e.g., during transportation in winter. The decrystallization by thermal effects would require a range of temperatures above 20 °C over a longer time period (unfreezing a bale of NR takes at least 4 days at 30 °C; a complete pallet might need some weeks!). Of course, partially crystallized NR has other mixing characteristics than non crystallized rubber. To consistently maintain the same quality, decrystallization must be assured. As already mentioned, this is normally done in the storing stage in a warm room.

A more effective approach to mastication is the so called "cold mastication". Here, a separate mixing step prior to the "usual" mixing procedure can be used. However, this method can lead to uncontrollable high current/power peaks in the mixer and to an inhomogeneous masticated batch.

As already mentioned, the viscosity of NR is highly dependent on temperature. Processors should be aware that if NR is masticated at high temperatures, shear forces could be too low in order to achieve good carbon black (or filler) incorporation and dispersion in the following mixing steps.

As the process of mastication is done in the absence of fillers, the mixer in this case might be underfilled and this in turn could result in less effective mixing. Processors should therefore perform some principal mastication tests before implementing such mixing steps. These can be done after the mastication step is completed and the polymer is dropped and then analysed in terms of homogeneity of viscosity and temperature. Normally, intermeshing mixers are better suited to masticate in a partially filled process than tangential ones.

Usually the mastication process as a first mixing step should be finished in a time frame of 45 seconds ($\pm$ 15). Processors should also make sure that the batch has seen enough rotations under the ram (at least 45) before other compound components are added.

It is recommended to use the highest possible rotor speed and a low specific ram force (25 N/cm^2). At this point in the process, the mixer has a low fill factor. The intake behavior for NR strongly depends on the surface- temperature of rotors and the mixing chamber. The use of higher temperature settings (minimum 40 °C) is recommended. The material behavior is also affected by the rubber temperature. In particular, the storage times of NR compounds between mixing steps have to be constant (minimum and maximum storage time). NR and NR blends exhibit a relaxation effect which has a big impact on the following processing steps.

The mastication effect can be influenced by special chemicals. Zinc oxide can be chosen to achieve faster tack to the surfaces of rotors and mixing chambers. To avoid excessive sticking to metal surfaces, zinc stearate, stearic acid, and sometimes waxes or low viscous polyethylene might be used [1]. If processors masticate natural rubber in a complete mastication step that is supported by aggressive chemicals (e.g., sulphur acid) it is important to use a special mixer design. In addition, the recipe should contain some parts of carbon black when the compound temperature reaches 130 °C. Above this temperature NR's adhesion to metal parts is higher in comparison to other polymers.

If peptizers are used, which are effective in accelerating the breakdown of rubber and therefore achieving fast mastication, it should be taken into account that sulphur or antioxidants may decrease the mastication effect. Peptizers are designed to saturate the free chain ends, which are produced by shear destructions of the molecular chains. As a result, the mastication effect is quite temperature dependent. If temperatures are low, the speed of the "peptizer reaction" is also low; however, at 120 – 130 °C shear forces in the mixer are getting too small to achieve mechanical degradation, therefore the mastication is also very slow at this temperature level.

As already mentioned, even higher mastication temperatures lead to thermal degradation and should be avoided. A maximum mastication speed can thus be realized in a temperature range from 90 – 110 °C.

Due to its high temperature dependence, the viscosity can drop quite quickly during the incorporation phase of fillers. As a result, NR shows a very pronounced second power maximum (see BIT in Chapter 2) after filler addition. The low viscosity at the end of the mixing cycle and the described high adhesion to metal surfaces can lead to a considerable flow in the direction of the dust stops at the end of the mixing cycle. Also, a penetration of the area between ram and feeding chute is likely to happen under those circumstances. Whenever possible, processors should try to avoid both effects by reducing the ram pressure at the end of the mixing procedure.

As natural rubber typically exhibits high elasticity, it can produce high pressures at the rotor flights. Consequently, strong ram movement can be seen, when NR is mixed.

A typical mixing process for high viscous NR is shown in Figure 3.1.

In step 1, polymer, zinc oxide, and peptizer are added to the mixer. The mixer (here 45L – intermeshing) is operated at high speed (90 rev/min) to achieve a high mastication effect. As a result, the mastication time itself is cut down to about 25 seconds.

The high temperature increase during this phase of mixing is acceptable because a lot of cold material is added in the following mixing steps and, as a result, the batch temperature is quickly reduced after the addition of the other components.

During the addition of fillers the speed is reduced to about 70 rev/min in order to guarantee a high power input. After the ram has reached its end position, the curing chemicals are added. In order to distribute these chemicals properly, the speed is once again reduced (to 45 rev/min). If the mixing conditions are kept constant, the viscosity level of the compound can be estimated on the basis of temperature and torque values (proportional to the amperes of the motor).

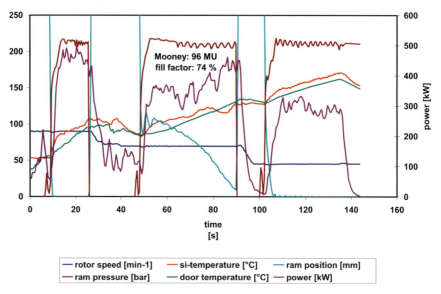

Figure 3.1 Mixing curve (fingerprint) for a NR masterbatch (NR 96 ME ML (1 + 4, 100 °C))

The high viscosity of NR allows to "overfill" the mixer to a certain level. It means that even when the ram is still about 10 to 25 mm away from its end position, an effective mixing at the ram bottom takes place.

Figure 3.2 also shows a typical low viscous NR mixing procedure. It can be seen that after the second ram lift the ram is not able to reach its final position, because

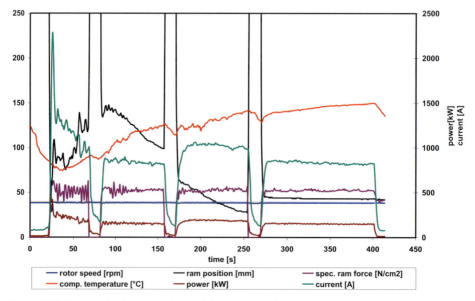

Figure 3.2 Fingerprint for a low viscous NR masterbatch

the low viscous compound is pressed into the gap between ram and chute in the last phases of mixing. The ram gets "jammed" in the chute and the compound in the mixing chamber is not affected by any ram pressure. Consequences of such "bad practice" are (at least) bad dispersion and potentially the long-term destruction of the walls of the feeding chute.

Depending on the height of the "jammed" ram, the temperature sensor is not able to monitor the proper compound temperature (bad flow around the sensor). In this case, the best option would be to lower the ram more slowly than normal (e. g., using a ram position control system). This phenomenon often occurs when mixing a second stage or a final stage of NR or NR blended compounds. Quite often processors are using the maximum ram pressure for all mixing steps. It is generally accepted that there is a correlation between ram pressure, shear force, and filler dispersion. As long as the BIT has not been reached, using the maximum available rotor speed and maximum ram pressure would be best. After dispersion is accomplished (as indicated by power or current peaks), the ram pressure can be lowered to a certain level in order to maximize lifetime of the mixer components and more importantly, to minimize problems such as a "jammed" ram. Usually the ram should "dance or oscillate" around its final position in the final mixing steps in all mixing stages.

NR is normally delivered in the form of bales. The mixing procedure is sensitive to the size of the dosed NR-lumps to the mixer (see also chapter two).

3.2 Ethylene Propylene Diene Rubber (EPDM)

EPDM is often processed in the form of polymer blends. Due to the strong polyolefin base (especially when the ethylene content is high) EPDM' are able to crystallize to a certain extent. To achieve specific final product properties, amorphous and crystalline types are often blended. As also shown in Chapter 6, fillers can have a tendency to move into a specific polymer. Due to this fact, often rather extended areas of unfilled and undispersed polymers can be found in these compounds because:

1. Crystallization of some areas of the "crystalline polymer". As the crystallization already takes place at room temperature and is only very slowly reversible, the dosing temperature should be at least above 15 °C. Since crystallinity tends to increase with a decrease in temperature, it is appropriate to provide adequate storage time under sufficient warmth (hot room as for NR) to minimize dispersion problems. Some processors use a special milling process before they add the crystalline EPDM types to the mixer. If the processors have to work with a

wider temperature range regarding the raw polymer, it is recommended to add the crystalline type together with the filler to achieve the highest possible shear and also to destroy the crystalline parts (polymer lumps).

2. Filler transport into only one phase of the polymer blend. This can be avoided by a good "pre-blending" technique of the respective polymers before filler dosing. At the end of this mixing phase a good and desirable level of dispersion of the different polymers within each other must be achieved, in which the domain sizes should be as small as possible. Processors should therefore evaluate the quality of this preblending step by dropping the polymers after the completion of the preblending phase using a visual inspection. However, as EPDM recipes have often a high filler content, it may be difficult to preblend effectively because the mixer is underfilled during this phase of mixing.

3. High viscosity differences between crystalline and amorphous polymer. These effects can make it very complicated to disperse the high viscous polymer in a low viscous matrix. In such a case, processors should try to create shear forces as high as possible at the beginning of mixing. To this end and in some instances, it may be desirable or necessary to load the compounding ingredients in one shot using "upside-down" (rubber put in last, after fillers, oil, and so forth)[1]. Under such conditions, the mixer is highly filled at the very beginning of the mixing cycle (but effect 2 has also to be taken into account!). Also, higher rotor speeds can help to achieve better polymer dispersion. A challenge is mixing compounds with a high polymer viscosity but with a low batch viscosity at the end of the mixing cycle. Sometimes, the best way to process these different polymers is the addition of the soft amorphous polymer later in the mixing cycle, even after the BIT.

Another unique effect of EPDM recipes is the effect of "black scorch". This means that at the end of the mixing cycle the compound viscosity starts to increase which results in a steady increase in the power demand of the mixer or a separate power peak. In further processing, scorched particles and surface defects can be found. To get "black scorch" no curatives, but rather EPDM and carbon black are necessary. In [2] this effect was investigated and it was shown that it is more pronounced for high ENB contents and high structured carbon blacks. It seems that bonding between the reactive surface areas on the carbon black side and some reactive ends on the polymer is responsible for this phenomenon.

The effect of black scorch can be suppressed by the addition of a very small amount of sulphur (about 0.1 phr) prior to filler incorporation. Figure 3.3 shows a typical one-stage mixing cycle for a recipe for technical rubber goods. At the beginning of the cycle, all compounding ingredients are loaded "Up-Side Down" into the machine. It can be seen that the power input is quite high, even at the start of the

---

[1] This means to dose the fillers first and the polymer(s) afterwards

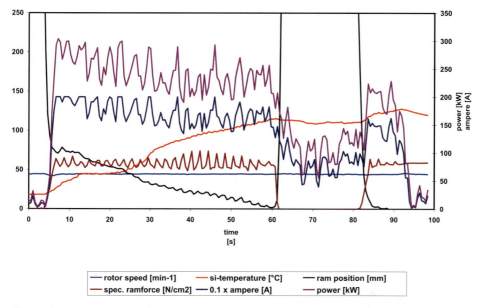

Figure 3.3 Up-side down mixing cycle for an EPDM compound

cycle. Thus, good polymer as well as good filler dispersion could be achieved. After a final ram venting, the compound was dropped at a total cycle time of only 95 seconds.

EPDM can be supplied in bale, granule and powder form. A special feature is a "friable bale". These bales can be disintegrated by small forces and lead to a quick "crumbling" of the material. Antitack agents are used to keep bales in a "friable" form and to prevent the adherence of granules. If particle sizes are small at the beginning of the cycle, it is important to achieve a quick power input (in general: higher fill factors, high speeds, "up-side-down", and so forth).

EPDM compounds are able to absorb a high content of oil in a short time. This is especially true if the compound temperature is elevated (> 80 °C). As already explained in Chapter 2, the carbon black itself can also absorb a high amount of oil. To achieve this, it ought not to be completely incorporated at the time of oil injection.

Processors have the possibility to either buy oil extended polymers or to incorporate the oils (free oils) during the mixing cycle.

The best way to incorporate oils is to use a torque related oil injection system. Additionally, to prevent the low viscous compound from creeping up into the feeding chute, the compound temperature should be kept constant after the oil was injected. A controlled setback of the ram in this phase (e.g., by a ram position control) would be more recommendable.

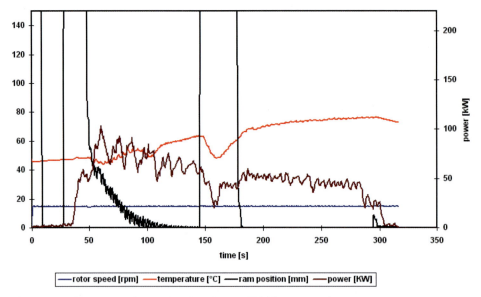

Figure 3.4 Power curve for an ultra low viscous EPDM compound

Figure 3.4 shows an example of a really soft compound. The viscosity of 7 Mooney Units was measured at ML 1+4 (120 °C). Compounds like this have to be mixed cold and the temperature settings are another main key for a good processing. The drop temperature was at 70 °C and the TCU set the rotors at 35 °C and the chamber at 45 °C. These compounds tend to stick to the rotors and to the rest of the mixing chamber. It is essential for a good quality that the mixer is completely emptied after the compound is dropped.

■ 3.3 Chloroprene Rubber (CR)

In principle, CR is able to react in a cyclic reaction. There are both sulphur and mercaptan modified types on the market, of which the latter are mostly used. Both exhibit a slow variation of their viscosity during storage and are able to crystallize. and their recipes contain slow, medium or fast crystallization rate types. As the "cyclization" goes on during the processing chain, temperatures should be kept as low as possible. In the mixer temperatures higher than 130 °C must be avoided and it is even recommended to stay below 110 °C.

CR is usually delivered in chips, which are approx. 75 mm long and contain anti-tackifiers to prevent the chips from sticking to each other and to guarantee quick power input into the compound.

Due to its chloride contents CR emits corrosive media. When mixing of CR compounds is suspended for a longer period of time (e.g., before a weekend), the mixer must be cleaned. This can be achieved by using specific cleaning batches with a saturated polymer.

CR exhibits the same constraints and limitations as other crystallizing polymers (e.g., NR). CR compounds can also crystallize. This could be suppressed by the addition of small amounts of BR (e.g., 5 phr). The crystallisation rate for CR-based compounds is generally lower compared to the raw polymer. CR shows only a limited ability to absorb oils; therefore, oil addition needs special attention.

Figure 3.5 shows an example of a CR compound. In this example, the aim was to keep the original recipe and to modify the process to achieve optimum dispersion. First, only the carbon black was added to the polymer. After the passage of the power peak (at 100 seconds) the oil was added in multiple injections (torque controlled). With this recipe – although the synthetic oil content was only small – it was not possible to inject the oil in one shot without a total "drowning" of the mixer. To control the torque it was important to add a frequency controller to the oil injection pump for the mixing line.

The amount of oil for other recipes at this mixing line was high and to reduce injection time, the flow rate for the mineral oil was set to a high level; however, this approach did not work for the CR compound. Using a high oil flow rate resulted in an uncontrollable oil addition step.

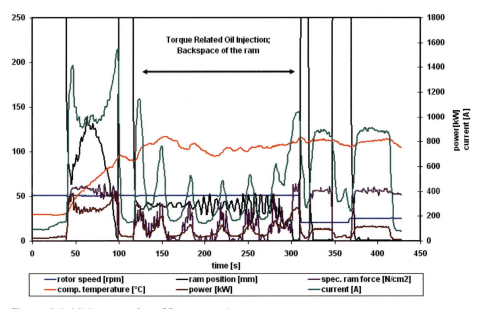

Figure 3.5 Mixing curve for a CR compound

To achieve good oil incorporation, the temperature should not be below 70 °C (100 °C is optimal). As viscosity decreases during oil incorporation, it is recommended to elevate the rotor speed during this phase.

In this processing example there was also the risk for parts of the low viscous compound or the oil to extremely contaminate the feeding chute. To avoid this, a controlled set back of the ram (about 10 – 20 mm) should be used during the oil addition. After the last oil addition, the processor has to define a temperature, current or power level as a "step switching" condition within his mixing cycle. In the step after this the ram can then be set back in the lowest position to mix most effectively and efficiently.

Modern control techniques allow combining a variety of controllers such as temperature controller, torque related oil injection controller, and ram distance controller. Most CR compounds can be mixed in a single stage process. If processors have to calender CR, it is possible to masticate CR (depending on CR-type) using the same formulation and procedure as already described for NR.

■ 3.4 Styrene Butadiene Rubber (SBR)

SBR is the main polymer used in the tire industry. There are two major types on the market: Emulsion polymerization ESBR and solution SBR (SSBR).

Depending on the "cis" and "trans" structure and with a narrow distribution of molecular weight of the polymer in SSBR, there are several possible advantages:

- better elasticity
- less heat build up
- less abrasion

Therefore SSBR is mostly used for tires. A typical recipe for treads could be:

SSBR/BR	52 %
Silica + silane	39 %
Antioxidants, ZnO, stearic acid	3 %
Oil	4 %
Curing system	2 %

High active fillers, sometimes together with a high structure, have to be incorporated to a high level. SBR is not as temperature sensitive as NR and shows a medium ability to absorb oil. Attention must be paid to oil addition to achieve a proper oil absorption by the fillers.

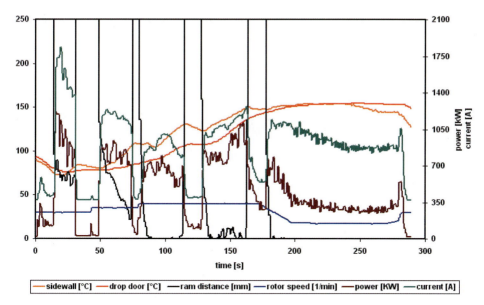

Figure 3.6 Power curve for a silica tread compound

Figure 3.6 shows an example of a 100% silica tread masterbatch compound for passenger cars processed on a 320 L machine. After blending the material with a quite good fill factor and several additives, (temperature-dependent) "silanization" takes place and thus it is important to keep a certain temperature range (see also Chapter 4). In this example, the temperature is kept at an elevated level of 155 °C.

Once silanization occurs, the process is kept at a constant temperature level for a certain number of revolutions. Because this process is a chemical reaction, it is important to process these kinds of compounds at the highest possible temperature. As a rough estimate, processors can calculate that an increase in mixing temperature of 10 °C will allow a reduction in reaction time by 50% (Arrhenius). It is very important to establish a small temperature distribution after the compound is dropped. Some rotors are not optimized in terms of maximum temperature homogenization. Therefore, processors have to compound these recipes at lower temperatures. Venting off by-products such as ethanol and water also has to be achieved during the silanization step.

When adding silica or other fillers at a higher level, processors should lower the ram with lower speed. If ram actions, such as ram lift or ram cleaning, are not used, some materials may be able to flow around the ram and stay there. To achieve the best dispersion and the required quality, it is important to ensure that all of the compounding ingredients are included in the mixing chamber. Measurements have shown that on a GK320 liter machine, up to 2.5 kg of fillers per batch can be

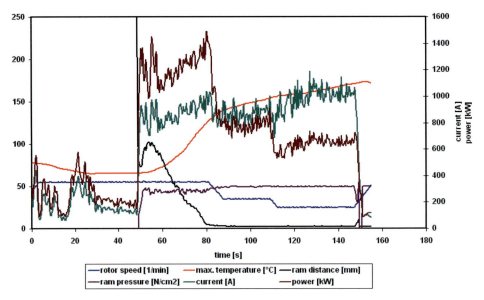

Figure 3.7 Fingerprint of a mixing cycle for ESBR blend

lost by aspiration. However, by slower movement of the ram during descend this lost can be reduced to only 200 g/batch.

Figure 3.7 shows mixing of a carbon black filled ESBR blend in a 320 liter machine.

Recipe:

ESBR (SBR 1712/SBR 1502)	59%
Carbon black (N220/N234)	35%
Small chemicals	3%
Oil	3%

Today, carbon blacks N330 or N339 with a higher structure are used.

Instead of oil extended SBR 1712 processors can use non-extended grades (cheaper polymers) with the equivalent amount of free oil being added during the process. Because of the amount of oil already included in the polymer, the raw viscosity of oil extended polymer is lower. This can lead to a more sophisticated mixing process to achieve good dispersion compared to a non-extended oil polymer type.

Adding certain amounts of oil later in the process could bring some benefits in terms of providing an active cooling process which can be used in process optimization, using a single stage process.

Old fashioned (conventional) processes consist of the following steps:

- Add polymers

- Mix for 45 seconds
- Add fillers (mainly carbon black)
- Mix until the compound reaches 90 °C
- Add oil
- Mix until the compound reaches 130 °C
- Lift the ram
- Mix until the compound reaches 150 °C
- Dump the compound.

Today modern control techniques are used to achieve better control during mixing.

Typical recommendations are:

- Blending of polymers as a first mixing step in order to achieve better filler incorporation into all polymers of the compound
- Mixing to a certain level of ram distance to retain enough open carbon black for oil absorption. Once this level is reached, start the oil injection.
- When there is no significant amount of free fillers left, lift the ram a few inches just before injecting the oil into the mixing chamber.

When discussing a single stage process and very short mixing cycles, processors have to keep in mind that the mixer needs a minimum number of revolutions to mix these components. Tangential type mixers are often equipped with a constant friction ratio for the rotor speed. Measurements over the years showed that batch-to-batch homogeneity and homogeneity within the batch are much higher when the best angle (rotor orientation) for an even rotor speed is used.

A typical recommendation is:

ZZ2 rotor geometry	330°
MD-SC rotor geometry	180°
Full four wing rotor geometry	90°

Some companies are producing SBR compounds with a low oil content. Here, processors should be aware of the operating conditions of the dust seals for these mixing lines. The allowed temperature depends strongly on the oil used for the dust stops. For these SBR-compounds it is not unusual to use up to 9.6 liter/hour for the sliding rings and 6.4 liter/hour for the annular gaps. In particular the amount of oil in the annular gap has to be correct to provide a certain slippage. At such high oil dosing, part of the lubricant might enter the mixing chamber. This is no problem as these compounds are able to absorb oil at higher temperatures. Therefore, depending on the recipe, processors should be able to use high amount of oils at the dust stops while producing this kind of compounds. As described above, really "dry" compounds can be produced with a back space of the ram.

3.5 Butadiene Rubber (BR)

BR is used as a blend partner for SBR in many tire applications. BR has a remarkable ability to absorb high filler and high oil contents. The viscosity of this polymer is much less temperature dependent than the viscosity of other polymers. At room temperature BR shows a tendency to "cold flow". This leads to the effect that the lower bales on a BR pallet are compacted, deformed, and therefore harder to dose than the bales higher on the pallet. Due to their sticky surface, BR bales are able to absorb almost any dirt particle in the raw material storage. They should consequently be stored in extraordinary clean conditions (and not on wood pallets).

At elevated mixing temperatures the viscosity of BR polymers does not drop dramatically and therefore approaches the viscosity levels of the other polymers in the recipe (normally SBRs).

3.6 (Acryl)Nitrile Butadiene Rubber NBR

Usually NBR polymers exhibit high viscosities, which causes very high current consumption during mixing. Figure 3.8 shows an example of a mixing curve (in the first mixing stage) for processing this type of compound (produced from a 45 liter mixer).

The maximum current is approx. 2500 A (the nominal current for that machine is 890 A). This means that the size of the drive train for such compounds is very important. It is not necessarily true that the materials with the highest Mooney viscosity are responsible for the highest loading demands on the drive train. Materials with "medium" viscosity (often NBR recipes) with high fill factors and short mixing times may pose the highest demands on the installed drive train, because Mooney viscosity is not always representative for the viscosity at the shear rates experienced during mixing.

Another result of mixing high viscosity compounds is the temperature development in the batch during the cycle. The compound temperature should be controlled intensively. Again, Fig. 3.8 shows that the compound temperature is reasonably controlled at 165 °C with the use of a temperature controller. Due to the good cooling capability of the intermeshing rotors, the temperature can be maintained by varying the rotor speed. The goal in this trial was to extend the cycle to reach a certain quality level. To reach a stabilized quality in production (with all the boundaries), it is necessary to use the total number of revolutions in the mixing step (not the time) as a trigger point for switching conditions when using a temperature controller.

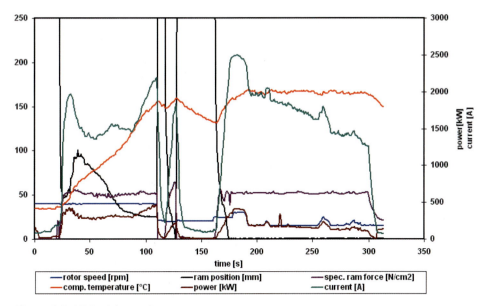

Figure 3.8 NBR mixing cycle

In terms of the final quality, it is debatable whether a fill factor of 65 % for the trial in question in Fig. 3.8 is optimum. A lower fill factor results in a faster ram seating during the dispersion phase (between 20 – 120 s) with the possibility of a higher number of total revolutions. Both factors have a considerable impact on productivity. It can be said that improvement in productivity was as a result of reduction in total cycle time.

The intake behavior of hard NBR compounds is sometimes very unfavorabled. The addition of NBR bales or NBR second stage masterbatches can result in a very high monitored compound temperature, because of the friction occuring between the rubber and the machine thermocouple. The operators should reduce the rotor speed during the addition of the rubber and use parameters other than temperature as a trigger point when using automation. Splitting the amount of rubber in two halves is another common practice to minimize this problem during feeding of masterbatches in production.

Depending on the rotor geometry, the intake behavior – as described above – is strongly dependent on the rotor speed used. The use of a lower rotor speeds can result in a quicker "drag-in" during this feeding phase. Because of the strong temperature increase during mixing, most compounders use bounded curing packages to distribute the chemicals in a short mixing time towards the end of the mixing cycle. As high shear forces are developed within the mixer, short mixing cycles can be realized.

3.7 Butyl Rubber

Butyl rubber is not compatible with other "diene" polymers because other polymers will absorb and react to a much larger extent with the reactive chemicals. For this reason, even small residuals from other polymers will lead to uncured areas if butyl rubber is not kept strictly separate from other polymers.

It is best to keep an individual mixing line solely dedicated to mixing butyl containing compounds. If not possible, the mixer must be cleaned intensively after every lot of Butyl compound being mixed. As a practical support, machinery suppliers are offering today the option of an automatically cleaning cycle including the dust seal areas.

As the viscosity of butyl rubber is low, the mixer should be filled as high as possible, with all compounding ingredients being loaded at the beginning of the mixing cycle. However, it is recommended that the oil is added shortly before the filler incorporation is completed.

3.8 Fluor Rubber

Fluor compounds are highly viscous. As they need no real mastication, "Up-Side Down" processes typically produce the best results. Figure 3.9 shows a typical mixing cycle. The mixer is run at quite low rotor speeds to avoid a too fast temperature increase. Also, the temperature control unit's (TCU's) settings should be set as low as possible.

As mineral fillers are usually used, some aspects of these materials may require special considerations. Mineral fillers tend to stick to metallic surfaces, if they are wet (high moisture content), therefore it is advisable to keep the moisture level as low as possible. To avoid condensation within the mixer, the temperature settings have to stay above a critical value. The ram in particular should be run at a higher temperature to keep it dry.

Figure 3.9 shows that the temperature increases fast, even at the comparably low rotor speeds. The ram position curve shows that the mixer has problems to ingest the highly viscous polymer. Ram seating took approx. 90 seconds. After the addition of the curing chemicals (rotor speed is reduced) the completly distributing of the components took approx. another 50 seconds. Most of the recipes were produced by an "Up-Side-Down" process. In general, these recipes do not contain active fillers; therefore reaching the desired level of dispersion does not pose an issue.

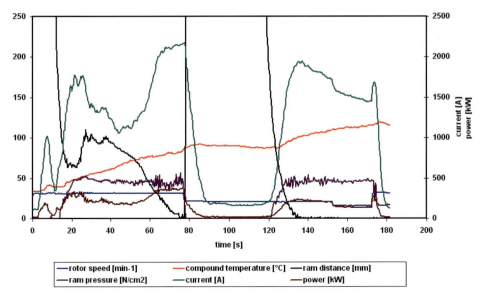

Figure 3.9 Fluor rubber mixing cycle

Quite often fill factors of 50 – 55 % are used for very expensive, highly viscous FKM types. Especially intermeshing mixers are able to produce the required quality level working with a fill factor in the above range. It has been shown that productivity, even running at a very low rotor speed and at lower fill factors, can be increased by up to 500 % compared to an open mill operation.

Customers are typically concerned about oil contamination resulting from the dust seals. This type of polymer must be processed with care; the mixer must be clean and free of oil and grease. A high level of cleanliness is important and therefore reducing the amount of oil for the dust seals is advisable.

∎ 3.9 Resins

If resins are components of a recipe, they need special attention. Quite often they melt only at temperatures above 60 °C, which also means they are rigid below this temperature. In the incorporation phase, they tend to stick to the cold surfaces of the mixer and the feeding chute, a problem that must be avoided. If the ram is lifted during a mixing step when resins are not completely dispersed in the batch, they tend to stick and immediately solidify at the walls of the chute. After a number of batches, serious contamination of the chute can be observed. This can quickly lead to very extensive damage to the feeding unit.

Resins such as HMMM, which are used in the remill or final stages of body compounds in the tire industry to get the required tackiness, have to be preheated to a temperature of approximately 60 – 70 °C before they are injected into the mixer. The higher temperature helps to reach a certain viscosity level, so that processors are able to dose the liquid material into the mixer via the oil injection valve. In addition, the incorporation of these resins in the compounds is much easier at this level of temperature. As for oils, injection of liquid materials can be done with ram down during a mixing step.

Other resins should be added to the mixer as early as possible. Cumaron resin for example, which has a melting point of 90 – 125 °C, can be added at the beginning of the mixing cycle. During the dispersion phase the resin is cut into very small pieces due to the high shear forces, resulting in a maximum surface melting the resin as fast as possible and this in turn results in an optimum distribution.

■ 3.10 General Considerations

Fill Factor
As shown in Fig. 3.10, the optimum fill factor for a compound is dependent on the (Mooney) viscosity. The displayed series is the result of many different trials using different mixers. It demonstrates that lower viscosities allow using higher fill factors.

Regarding the fill factor, the following points should be considered:

- The highest fill factor does not necessarily lead to the highest throughput! However, lower fill factors could lead to higher productivity as they allow in many cases more favorable process conditions than higher rotor speed and a better temperature control. It should be stressed here that the key intention always is to achieve the desired level of dispersion and distribution quality in the mixer.

- The "total revolutions" under the ram should also be taken into account. Higher speeds (achieved, e.g., by lower fill factors) could lead to more revolutions and thus to a better distribution, or – in the case of silica compounds – to a better devolatisation.

- The fill factor should be optimal for *every* step of the mixing cycle. In practice, this is not always possible and therefore compromises for certain mixing phases have to be made.

- Observing movement or "playing" of the ram at the end of the mixing cycle is a good method for evaluating the fill factor for most recipes. However, we have observed compounds (especially for tangential machines), where the fill factor displayed a perfect "playing" ram, and yet turned out to be too high.

3.10 General Considerations

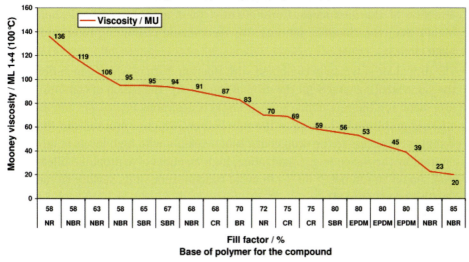

Figure 3.10 Relationship between fill factor and Mooney viscosity

- Recipes with high filler content allow for a relatively higher fill factor to be used for the mixer compared to recipes at the same viscosity level but with lower filler concentrations.

The fill factor for all recipes ranges between 0.5 and 0.92. The batch weight can be obtained from the following calculation:

Intermix 190E = 190 liter empty volume,

Density of mixed compound = 1.2 g/cm³

Fill factor = 0.7

Batch weight = 190 · 1.2 · 0.7 = 159.6 kg

Temperature Settings

As all temperature controlled parts of the mixer have different functions, the water temperature in these parts must not be set to the same temperature levels. If possible, the settings should be different for the various parts of the mixer.

- To achieve a high conveying capacity of the rotor flights (good distributive mixing), the material should stick to the chamber walls and glide on the rotor surfaces. The latter does not support ingestion of the material. To support the different dragging/conveying characteristics, the rotor temperature should be set at a higher temperature level than the chambers. In practice, a temperature difference of 10 °C is recommended.

- The wear plates should also support the gliding of the material in order to prevent sticking and to minimize drag flow (caused by the "Weissenberger effect") in the direction of the dust stops. Consequently, their temperature level should be higher than the chamber settings. We recommend combining rotor and wear plate circles.
- The drop door is typically combined with the ram, both of which can suffer from the same sticking effects (at the drop of the batch). The option of combining the drop door and the ram in one zone or circle is desirable. Processors can analyze sticking problems by looking at the ram bottom (which is much easier than the complicated access to the drop door). When dealing with high levels of moisture emitted by the recipe (e. g., silanization; high amounts of hydrophilic fillers), the ram should be ran hot (> 80 °C) to avoid condensation at its surface (it means a fourth cycle is needed).

Ram Pressure

As already described earlier, high ram pressure should be used until the final dispersion quality is reached. Dispersion quality increases when using a higher ram pressure during the mixing phase/mixing step. As soon as the BIT is reached during the process, the ram pressure can be recuded. Finally, before dropping the compound, depending on the rotor geometry, it is important to keep the ram at its final position.

Using maximum ram pressure for all mixing phases in all mixing stages results in increased wear and tear. However, it is important to apply a constant ram pressure during the dispersion phase. Without closed loop control of the ram pressure, inconsistency in quality and batch-to-batch variation may be encountered. While producing second or final stage compounds, a reduction in ram pressure is possible while maintaining quality and, at the same time, increasing the lifetime of the mixer (less wear).

Ram Position

In most cases, the position of the ram indicates whether the used batch weight is correct. If the ram does not touch the final position during the mixing process, this is an indication that the batch weight is too high! If the ram has not reached its final position, the ram should neither be lifted for cleaning nor for turning the compound above the rotors. While tangential type mixers drag in the material immediately and start mixing from the top of the drop door (bottom up), the intermeshing type mixers start from the top of the rotors (top down). In tangential type mixers the ram will reach its final position after a few seconds and move around this position. When using intermeshing mixers, the intake behavior (ram movement) depends mostly on rotor geometry and ram pressure. At high fill factors, the ram can touch its final position just a few seconds before the compound is mixed and ready to drop. Both systems are totally different in terms of their mixing behavior and process optimization.

In the past, it was only possible to set a certain pressure and to try to control the set points. Today, it is possible to define a master curve and then the ram is forced to follow the programmed curve. There are several advantages to mixing using this tool:

- Fibres can be added and incorporated under tightly controlled conditions. When using intermeshing mixers, this is a big advantage and beneficial as these mixers are mixing top down. Fiber bales can be added and kept at a certain length (much longer compared to other processing concepts such as co-rotating twin screw extruders).
- For processors it is only important that the ram moves around its final position before dropping the compound. At this stage, the ram pressure can be lowered. This can be done automatically by the control system.
- Wear and tear of the mixer (especially chute and dust seals) will be reduced when using such a system.
- Programming a sinus curve for final mixing or for distribution steps eliminates the need for lift steps and results in a higher output.
- The negative effect of adding different lump sizes of polymers (see Chapter 2) can be eliminated completely.
- The use of different ram speeds when lowering the ram will keep the material in the mixing chamber. The material will not bypass the ram and stay on top of it. Especially when fillers, such as silica, or free flowing materials, such as pyrogene silica, are used. In particular it is important for the dispersion level and quality of the mix to have all of compounding ingredients present in the mixing chamber when starting the mixing process, which results in the highest possible shear forces.
- Processors could work with a controlled "set back" of the ram. This means the "Zero position" is shifted a certain way up into the chute. By using the "set back", control of the ram position (elevation in the throat) can be defined separately for each recipe.

Rotor Speed

It is recommended to start the mixing process with the highest possible rotor speed. Cold added materials, such as polymers and fillers, result in the highest possible shear forces when a high rotor speed is used. Later in the cycle it is common to slow down the rotor speed to reduce the rate of temperature increase in the mix. The main aim is to prevent an increase in compound temperature that could possibly lead to polymer degradation. The optimum rotor speed depends on the design of the installed rotors. Rotor geometries are optimized to achieve maximum dispersion or a maximum distribution. For example, ZZ2 rotors are designed to reach maximum distribution at rotor speed rates up to 50 rpm. These rotors allow mixing of curing agents and provide a much better distribution while reducing the energy input by 30 % compared to Full-Four-Wing rotors. If open wing rotors such as ZZ2 are used for dispersion phases, it should be possible to run the mixer at a higher rotor speed than a Full-Four-Wing (e. g., 70 rpm).

When using intermeshing systems, processors have to keep in mind that at high speeds and due to the smaller clearances between the working surfaces of the mixer (like a mill), the intake (drag in) of the material between the rotors is minimized. Therefore it is recommended to reduce the rotor speed for a short time during this phase. This allows for an optimized feeding sequence especially of the masterbatch from earlier mixing stages. When compounds like these are kept too long on top of the rotors (tangential or intermeshing, depending on batch weight), the compound viscosity is expected to be high in this region, whereas the viscosity in the mixing chamber is low (higher temperature). If different areas of the mix exhibit different viscosities, it will have a negative impact on the following processing steps.

Power Demand

The required power demand depends on:

- Fill factor
- Compound temperature
- Mixing time
- Viscosity
- Rotor speed

The drive train settings have at least two limitations. First, they have a maximum power level. As soon as the maximum current is reached during processing – e. g., at the highest rotor speed needed for a certain time span – rotor speed will be reduced together with a reduction in ram pressure. Second, during the operation the systems will be warmed up (e. g., frequency controller). If critical temperatures are reached, the motors will switch off which can lead to severe practical problems (e. g., a mixer filled with cold cured rubber). Modern technology can monitor both limitations and processors have the option to choose fill factors or mixing steps to prevent an undesired motor-stop.

In particular, if more than one mixing line is run simultaneously, the addition of high power demands can lead to an extraordinary load on the power supply and to high costs to serve these power peaks. As a consequence, it is advisable to plan the production with respect to the power demand of the recipes and to avoid the simultaneous production of high power demanding compounds.

Running medium viscosity compounds – with a high fill factor at medium speed – requires maximum loading. When using lower rotor speeds (instead of a high rotor speed), the current peak will be kept over a longer period of time during the mixing step. This of course could result in problems with temperature increases in the frequency controller of the main drive. In such cases it should be analyzed whether a lower fill factor in combination with a higher rotor speed (shorter timespan for the powerpeak) is more advisable.

■ References

[1]   Schnetger, J.: Kautschukverarbeitung, Vogel Verlag Würzburg 1998
[2]   Daniel, C., Pillow, J. G.: Black Scorch in EPDM Compounds, 99 Rubber Conference, Manchester June 1999

4 Internal Mixer – a Reaction Vessel

Oliver Klockmann

An essential condition for optimum reinforcing properties is good filler dispersion in the rubber compound as this allows best interactions between rubber and fillers. Reinforcing properties have a fundamental influence on the physical properties of the compound and determine the usage properties of rubber ware.

For decades, carbon blacks were the only additives used to increase reinforcement properties; however, today it is possible to achieve more individual characteristics by using a different type of filler: precipitated silica. Compared to carbon black, precipitated silica alone reinforces significantly less, due to its high polarity and because rubber and silica are not compatible due to their different polarity. Therefore, bi-functional organic silica compounds (organosilanes) are used, acting as coupling agents between rubber and silica, building a bridge between the two materials. The use of bifunctional organosilanes in rubber applications in the 1970s marks the beginning of chemical reinforcement [1–3]. Then, covalent bonds between fillers and rubber were introduced. This opened completely new application fields for mineral fillers, particularly for highly active precipitated silica. In all fields, but especially in the tire industry, an evolution started that is far from being completed. Although the silica-silane technology was first introduced to the tire industry in the early 1990s [4–7], only a decade later, almost all passenger cars in Europe were equipped with tires whose treads contain silica and silane. The reason was the particular property profile that can be achieved thanks to the so-called "Green Tire" technology. Compared to the formerly used carbon black filled systems, the "Magic Triangle" could be widened (Fig. 4.1). By then the assertion that important properties, such as rolling resistance, wet grip, and abrasion resistance could only be improved by curtailing the others – the improvement of a property resulting in the deterioration of another – was not valid anymore. Silica-silane technology allowed a significant improvement of rolling resistance and wet grip without having a negative influence on abrasion resistance and hence on the service life of tires.

Rolling resistance was reduced by approximately 25% and the fuel consumption by approximately 5%. This new tire generation showed much better performance in all important properties. However, it came with the price of deteriorated processing properties in comparison to conventional carbon black systems. In addition to dis-

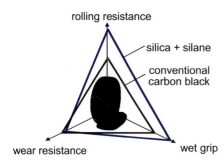

Figure 4.1 Widening of the magic triangle thanks to the silica-silane technology

persion and distribution of the fillers and further ingredients, a chemical process was now crucial for the mixing process. Hence it was and is still indispensable to investigate the influence of the various factors on the reaction, such as temperature, time, concentration, and catalysis and to use the results to optimize compound manufacturing. The internal mixer turns out to be a chemical reactor.

To fulfil the task of bonding the fillers to the rubber, the bifunctional organosilanes use two reactive groups in their molecule [8–11]. The trialkoxy silyl group is able to react with the silanol groups of the silica, building stable siloxane bonds (filler modification). This should proceed during the compound manufacturing and is accompanied by a cleavage of ethanol [12]. The second group, the rubber active one, reacts with the polymer during vulcanization and develops covalent filler-rubber-bonds. These chemical bonds are responsible for the high reinforcing potential of the silica-silane filler system [13, 15].

To achieve best processing and performance properties of the vulcanizate it is important to make sure that during the compounding process the coupling agents and the silica are mixed to a very high degree. At the same time, a reaction with the polymer matrix should be prevented. These chemical reactions have to be carefully controlled.

In the following we will explain the reaction of the organosilanes with the silica during the mixing process and point out the basic conditions for mixing of compounds with a high load of silica.

■ 4.1 The Silica Network

Due to the polar surface of precipitated silica, there is only a weak interaction potential with rubber, but a strong silica-silica interaction via hydrogen bonding. This leads to the development of a strong silica-silica network and therefore to a strong Payne effect. This filler network is further responsible for an increased Mooney viscosity, a reduced incubation time during vulcanization and a low degree

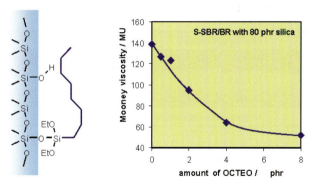

Figure 4.2 Hydrophobation of the silica surface

of reinforcement. One of the major tasks of rubber silanes is to overcome this tendency of silica to agglomerate by hydrophobation of the silica surface. Monofunctional alkyl silanes are used to study the effect of hydrophobation. They only bear a silica-active site and are not able to react with the polymer. By reaction with the silanol groups on the silica surface, the number of polar -OH groups is diminished, the silica surface becomes hydrophobic. Depending on the alkyl chain length of the investigated alkyl silane, further -OH groups can be shielded and the polar character of the silica surface is further reduced (Fig. 4.2). As a consequence, silica agglomeration is strongly inhibited and, e.g., the Mooney viscosity is strongly reduced. As shown in Fig. 4.2, only small amounts of an alkyl silane, such as octyl triethoxysilane (OCTEO) are necessary to induce a strong decrease in viscosity.

As can be easily seen in Fig. 4.2, a "full silica compound" (80 phr HD silica Ultrasil® 7000 GR in an S-SBR/BR blend, commonly used in passenger car tire tread compounds) without any silane would not be processable. In an industrial process, Mooney viscosities beyond 100 MU after the final mixing stage would result in serious problems regarding the subsequent steps such as extrusion. Therefore, silanes act as processing aids during processing. The silica network is partly broken down. Both the Payne effect and the Mooney viscosity are reduced. Further examination of the Payne effect shows the collapse of the silica network caused by hydrophobation [16].

■ 4.2 Influence of Mixing Time and Temperature on Hydrophobation

Together with the temperature profile in the internal mixer, the mixing time is an important parameter for the chemical reaction of the silane with the hydroxyl groups on the surface of the silica. In order to scale the state of hydrophobation

Table 4.1 Mixing Process for Measuring of Silanization Reaction

Mixing time (s)	Mixing procedure
0–1	Addition of polymer
1–2	½ silica, ½ Si 69®, ZnO, stearic acid, plasticizer
2–3	½ silica, ½ Si 69®, wax, 6PPD
3–15	Mix and take the sample out

after different mixing times, the hydrophobation grade of the silica was measured after 3 minutes every minute by means of RPA-measurements. The recipe corresponds to compound I (1st stage) in Table 4.1, using the highly dispersible silica Ultrasil® 7000 GR as the silica and the polysulfane Si 69® as the silane component.

The Rubber Process Analyzer (RPA) allows a detailed investigation of the degree of hydrophobation in the internal mixer. By means of shear the degradation of the filler network is investigated in relation to the mixing time. As expected, at the beginning of the mixing process a pronounced filler network can be observed at low shear rates (Payne effect), which disappears more and more due to hydrophobation with the silane. This reduction of the Payne effect occurs very slowly at low mixing temperatures (140 °C; Fig. 4.3 on the left) and the final value is first reached after 10 minutes, whereas at 155 °C mixing temperature (Fig. 4.3 on the right side), the hydrophobation reaction is already completed after 8 minutes and a low Payne effect is observed.

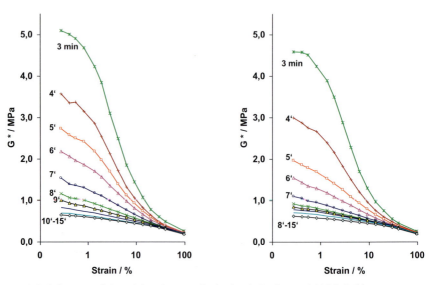

Figure 4.3 Influence of the mixing time on the hydrophobation at 140 °C (left) and 155 °C (right)

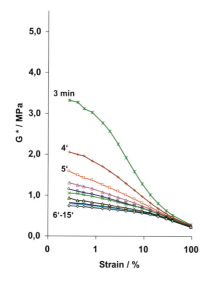

Figure 4.4 Influence of the mixing time on the hydrophobation at 170 °C

At higher mixing temperatures (170 °C, Fig. 4.4) the silanization reaction is the fastest, as expected. However, this temperature is not within the allowable processing frame for polysulfide silane Si 69®. A pre-scorch, which will be described later, occurs and would generate problems regarding processability and the in-rubber properties.

The silica-silane reaction is a chemical reaction that occurs during mixing. Therefore, control of temperature and time plays a crucial role. In the following, the chemical aspects of this reaction will be investigated in more detail.

4.3 Chemistry of the Silica-Silane Reaction

Coupling agents such as Si 69® dispose of triethoxysilyl groups, which can react with the silanol groups at the surface of the silica. To describe the mechanism of the reaction silica/Si 69®, investigations were performed in model compounds, such as decane, using ^{13}C and ^{29}Si solid NMR spectroscopy and kinetic studies [17, 18]. The investigations led to the reaction model shown in Fig. 4.5. This model subdivides the whole reaction in a primary and a secondary reaction.

The primary reaction is the reaction of the silane with the silanol groups of the silica, i.e., its coupling on the surface of the filler. Only catalytic quantities of water are required for this purpose. As shown in Fig. 4.6, the reaction rate at humidity levels higher than 6% is not increased. The investigations were performed at 50 °C because at this temperature only the primary reaction occurs, as it has a tenfold higher rate constant compared to the secondary reaction.

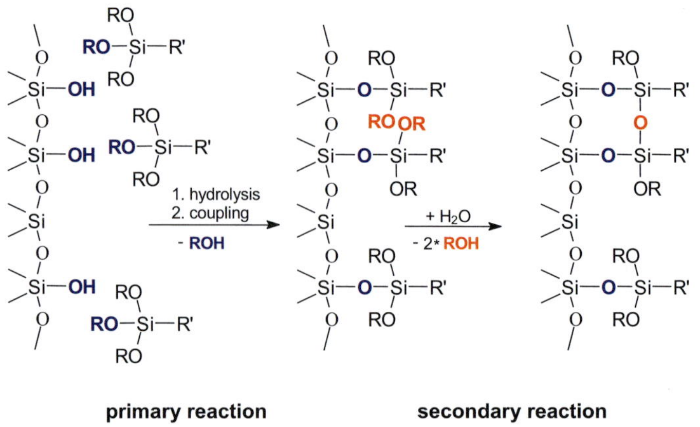

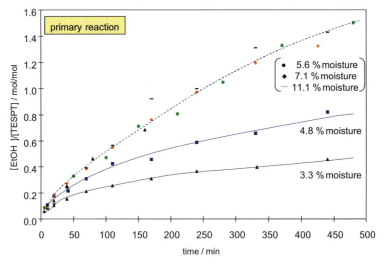

Figure 4.5 Overall reaction model of the silanization reaction of Si 69®

Figure 4.6 Reaction of the silica with Si 69® in decane at 50 °C, depending on the moisture content

The secondary reaction is the sum of numerous subsequent reactions and is defined as the condensation reaction of the silane molecules already bonded to the silica with further cleavage of ethanol (crosslinking). Unlike for the primary reaction, a hydrolysis step and therefore the presence of water is needed before this condensation reaction occurs due to mechanistic reasons. Hence, the yield and the reaction speed increase with rising moisture content, as depicted in Fig. 4.7.

During the primary reaction only two mol ethanol per mol Si 69® are released, whereas up to 6 mol ethanol per mol silane may be released due to the subsequent secondary reaction. Due to its low boiling point, alcohol is classified as a volatile

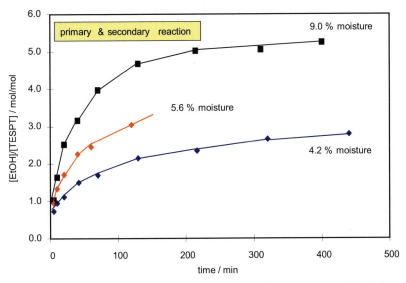

Figure 4.7 Reaction of the silica ULTRASIL® VN 3 with Si 69® in decane at 140 °C, depending on the water content

organic compound (VOC). This has several consequences. On the one hand, the mixing routine must guarantee that the ethanol can be released out of the compound – otherwise the reaction would be slowed down [19] and in addition, undesirable porosity could occur later during extrusion. Thus appropriate ventilation of the internal mixer is a precondition. On the other hand, the release of VOC into the environment (depending on the quantity of silane used) has to be controlled, e.g., by an incinerator. Releasing VOC out of the end product may also cause regulatory and other issues today. By ensuring a completed secondary reaction during the mixing process, the release of VOC, which in other cases would be emitted during further processing, can be preponed to a certain extent. As shown in Fig. 4.7, the moisture of the silica is also relevant. The managing of temperature and time offers further possibilities to change reaction characteristics; however, it is limited with respect to the used silane.

■ 4.4 Temperature Limits

The sulphur function of Si 69®, with a polysulfane distribution ranging from S_2 to S_{10}, is able to react thermally with rubber, even without any additional vulcanization agent [2, 13, 15]. This thermal crosslinking occurs at high temperatures, with slow linking speed and low coupling yield. If pre-scorch occurs due to too high temperatures, particularly during the dispersion phase of the filler, the viscosity of the compound increases, which leads to difficulties in compounding. Thermal pre-scorch should be avoided by any means.

Pre-scorch in the rubber compound can be investigated very easily using a vulcameter at high temperatures. For this purpose, rubber compounds were prepared with silane Si 69® (with an average sulphur chain length of 3.75), and with disulfide silanes with different average sulphur chain lengths, at moderate temperatures, respectively. The torque values of the unaccelerated raw compounds were measured at 180 °C (Fig. 4.8). Sulphur silanes with average sulphur chain lengths of more than 2.0, which also exhibit a share of tri- and higher sulfanes, already react in unaccelerated compounds. The pure disulfide sulfane with a chain length of 2.0 cannot react with the rubber and is thus depicted by the line at the bottom of the graph.

This possible thermal pre-scorch leads to restrictions during processing, especially regarding temperature. In order to investigate the influence of the dump temperature on the in-rubber properties, a well-known S-SBR/BR compound (Table 4.1) was selected and examined in a temperature range from 120 to 200 °C. The temperature in the lab kneader was adjusted by varying the rotor speed and the cooling temperature of the mixing chamber. In addition to Si 69®, commercially available disulfide silanes Si 75 (average sulphur chain length of 2.35) and Si 266 (average sulphur chain length of 2.2) were used.

The investigations showed that the viscosity was at a minimum in the range of 145 to 155 °C (Fig. 4.9). All silanes exhibit an increase of viscosity at decreasing temperatures. Too low temperatures have a negative effect on the quality of the silica hydrophobation and the silanization yields become too low. The increase in Mooney viscosity with temperatures above 150 °C and 165 °C, respectively is mainly due

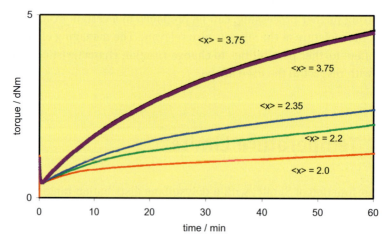

Figure 4.8 Torque increase of the unaccelerated compounds with sulphur silanes of various sulphur chain lengths <x> depending on the time

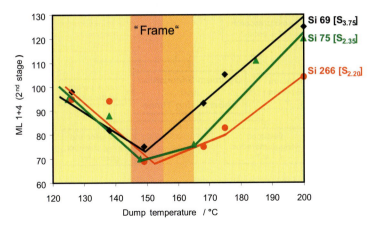

Figure 4.9 Dump temperature vs. Mooney viscosity

to reactions of the polysulfane groups with the polymer (pre-scorch). As can also be seen in Fig. 4.9, low Mooney viscosity depends on the sulfur silane used. It is possible to define "processing temperature ranges": in the case of Si 69®, the mixing temperature should range from 145 to 155 °C, whereas temperatures up to 165 °C can be tolerated using disulfane silanes such as Si 75 or Si 266.

■ 4.5 Summary and Consequences

The dispersion and the distribution of the filler in the rubber matrix is a basic process which takes place in the internal mixer. The interactions between rubber and filler determine the reinforcement properties and ultimately the usage properties of the elastomer and the compound, respectively. The dispersion has always been an important factor determining the compound quality. Hence, the mixing unit and mixing procedure aimed for a good dispersion, achieved under cost-effective conditions (short mixing times). Now, the use of silica silane technology is shifting these requirements. Due to the reaction of trialkoxysilyl groups with the silanol groups of the silica during manufacturing, not only the degree of dispersion but also a chemical process plays a decisive part. To obtain an optimum hydrophobation, the silanization speed (influenced by temperature, time, transport processes) must be adjusted to the dispersion process.

Figure 4.10 is summing up the various influences that have to be taken into account in the mixing process [15].

The mixing process has to comply with requests that are partly contradictory; for example, to prevent pre-scorch, the mixing temperature must remain low because low temperatures and short mixing times impede an early vulcanization. But at the

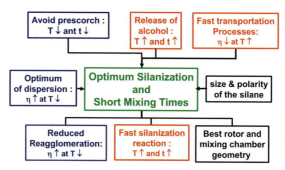

Figure 4.10 Influence of the silanization on the mixing process

same time, it is advantageous to keep both the temperature and the mixing time on a high level while venting the chamber to eliminate the released ethanol and the residual moisture. The transport of the silane to the silica surface plays an important part and is promoted by high temperatures, while this transport process as well as the silanization reaction also depends on the type of silane used. The geometry of the rotors and the mixing conditions should provide optimum support for the filler distribution and dispersion, the transport process, and the silanization. A purposeful temperature management is thus a precondition. The chemical process on the surface, the silica-silane reaction, is accelerated by high temperatures, whereas high shear rates and low temperatures are advisable to guarantee best filler dispersion and suppression of reagglomeration.

It can be stated that the requests on the mixing process are manifold. Homogenizing the compound is no longer sufficient. A controlled chemical reaction must be achieved and therefore the internal mixer turns out to be a reaction vessel. Due to partly conflicting requests, no general recommendation can be given. In fact, the mixing process of each silica compound has to be adjusted and optimized individually.

■ References

[1] Schwaber, D. M., Rodriguez, F.: *Rubber and Plastics Age*, **48** (1967), p. 1081
[2] Wolff, S.: *Kautsch. Gummi Kunstst.*, **30** (1977), p. 516
[3] Wolff, S.: *Kautsch. Gummi Kunstst.*, **32** (1979), p. 312
[4] Patent EP 0 501 227, US 5.227.425
[5] Trono, A.: TyreTech.'92, Paris/Frace (1992)
[6] Agostini, G., Bergh, J., Materne, Th.: Akron Rubber Group, Akron, Ohio/USA, Oct. 1994

[7] Marwede, G. W., Eisele, U. G., Sumner, A. J. M.: ACS, Cleveland, Ohio/USA, Oct. 1995
[8] Ranney, M. W., Pagano, C. A.: *Rubber Chem. Technol.* **44** (1971), p. 1080
[9] Thurn, F., Wolff, S.: *Kautsch. Gummi Kunstst.* **28** (1975), p. 733
[10] Wolff, S., Wang, M.-J., Tan, E.-H.: *Kautsch. Gummi Kunstst.* **47** (1994), p. 102
[11] Görl, U.: *Kautsch. Gummi Kunstst.* **51** (1998), p. 416
[12] Kiselev, A. V., Lygin, V. J.: Infrared Spectra of Surface Compounds. Wiley, New York/USA, 1975
[13] Görl, U., Münzenberg, J., Luginsland, D., Müller, A.: *Kautsch. Gummi Kunstst.* **52** (1999), p. 588
[14] Klockmann, O., Hasse, A.: ASC, San Antonio Texas/USA, May 2005
[15] Luginsland, H.-D.: A review on the chemistry and the reinforcement of the silica-silane filler system for rubber applications, Shaker, Aachen, Germany (2002)
[16] Fröhlich J., Luginsland, H.-D.: *Rubberworld* **28** (2001), p. 244
[17] Görl, U., Hunsche, A., Koban, H. G., Lehmann, Th.: *Kautsch. Gummi Kunstst.* **51** (1998), p. 525
[18] Hunsche, A., et al.: *Kautsch. Gummi Kunstst.* **50** (1997), p. 881
[19] Dierkes, W., Noordermeer, J. W. M., Kelting, K.-U., Limper, A.: *Rubber World* **229** (2004), p. 33

5 Effect of Process Parameters on Product Properties

Dr.-Ing. Harald Keuter (Section 5.1 to 5.7)
Dr.-Ing. Peter Ryzko (Section 5.8 to 5.13)

■ 5.1 Introduction

The rubber processing industry today is confronted with increasing economic pressures as well as increasing qualitative demands on the products it manufactures [1, 2]. Consequently, in the manufacture of rubber products it is necessary to continuously open up new potentials to meet these requirements.

Rubber products consist of a large number of different raw materials. Depending on the product requirements, customized formulations are developed from these raw materials. When the materials have been weighed out, compounds are produced. For this purpose mainly so-called ram kneaders or just "kneaders" are used. The compounds must then pass a quality control inspection before being further processed into end products.

Currently, quality control inspections in the mixing room identify only approx. 1 % of the compounds as rejects. At the end of the process chain there are often much higher reject rates, sometimes as high as 30 %. One problem is the fact that in the case of rubber products, rejects are very expensive because of the high costs of raw materials and in many cases the rejected end products cannot be reused. These facts are conflicting with the previously defined demands placed on rubber products.

One significant reason for the sometimes high rate of rejects certainly is the complexity of the material "rubber". Formulations for these products generally comprise up to 10 different raw materials. The condition of the raw materials can be very different: from dispersed bulk materials, fluids and wax-like substances, to high viscous polymers. In addition, every raw material is specified with different quality parameters. Consequently, the manufacturer is confronted with an almost immeasurable number of influencing factors that may have an effect on the quality of compounds and end products.

This multi-component mixture has to be mixed in the kneader in a reproducible manner until it is homogeneous. Sometimes the individual quality parameters may influence one another [3]. In addition, many plants process a large number of different formulations – up to 1000 in some cases [4]. Therefore, rubber processors generally regard the manufacture of customized compounds to be their core competency and operate their own mixing room.

Hence, certain facts would suggest that the discrepancy between the rejects detected in the mixing room and those in the end products can, in many cases, be attributed to the inadequate mixing quality not recognized during quality control in the mixing room. When the compounds are further processed, rejects are produced in the start-up and shut-down processes, in assembling, and during the coating or jointing processes, which is in many cases unavoidable. It can be assumed that these rejects make up 60 – 70 % of the entire reject rate. Consequently, an approach to reduce the sometimes high reject rates ought be a systematic examination of quality influences in the mixing room in regard to their effects on the quality of end products. If these relationships were better understood, there would be a chance of reducing the reject rates in the processing of rubber products. Forecasting the quality of end products with the help of quality data from the mixing room is very important. For this purpose, testing methods for characterizing compounds are required which are able to create correlations between the qualities of the compounds and those of the end products.

When we analyze the compounding process in the mixing room, we can define three areas that may influence the quality of rubber products: the raw materials and fluctuations in specified quality parameters, storage of raw materials and their transport to the kneader, and the manufacture of the compound itself.

So the aim of this chapter is to analyze the before-mentioned effects and to suggest a concept for improving the quality in the mixing room. It must also be discussed here to which extent control concepts can be successful and which contribution can be made by improved quality testing methods.

5.1.2 Quality Parameters of Raw Material

As already mentioned, the raw materials used in the mixing room are numerous. So it is hardly possible to give a complete summary of all possible parameters of each raw material. This article shall deal with examples of the quantitatively important parameters polymer and carbon black.

5.1.2.1 Quality Parameters of EPDM Polymers

Because of the importance of EPDM in the technical rubber goods industry, its quality parameters will be analyzed here exemplarily. EPDM polymers are characterized with the help of the following quality parameters: Mooney viscosity, ethylene or propylene concentration, type and quantity of the diene monomer, and

Figure 5.1 Mooney viscosity of an EPDM as function of the delivery batch

type and quantity of the extender oil [5]. The properties of the compound and the corresponding vulcanisates are directly affected by the polymer structure, namely: molecular weight, molecular weight distribution, composition, crystallinity, distribution of the monomers within the polymer chain, and proportion of long-chain branches [6].

If these parameters are analyzed from delivery to delivery, it turns out that there are only minor variations. As an example, Fig. 5.1 shows the Mooney viscosity of an EPDM polymer as a function of the delivery batch. The only quality parameter not specified, or specified only in an imprecise manner, is the proportion of long-chain branches. The effect of long chain branching on the process was described in [7, 8], another method for characterization, the so-called $\Delta\delta$ Parameter, in [9].

5.1.2.2 Quality Parameters of Carbon Black

Carbon blacks are very complex raw materials, which are also specified by a range of quality parameters. Carbon black is the most commonly used filler and it may have both pure filling and reinforcing properties. Usually, carbon blacks are specified by the following quality parameters: DBP absorption, CTAB adsorption, pellet hardness, fines content before delivery, grit, and bulk density. The specifications vary slightly from supplier to supplier and depending on the rubber processor and on the bilateral agreements between supplier and customer. The following will describe the significance of important quality parameters in more detail.

After an analysis of the distribution of the quality parameters defined in carbon black certificates during a one-year inspection period, only the pellet hardness proved to be significantly different from delivery to delivery. All other quality parameters showed very little distribution relative to the defined specification limits.

In the following we will explain the fundamental properties of the raw material carbon black, which are necessary in order to understand the effects of its quality parameters on downstream processes and/or product quality. 98% of carbon blacks used in the rubber industry are manufactured today using the so-called furnace process, in which hydrocarbons are incompletely incinerated or thermally split [10]. During the first stage of the formation of carbon blacks, highly viscous drops or solid particles form with a more or less spherical shape [11, 12]. As they grow together, three-dimensional branched structures evolve, which are described as aggregates. These represent the smallest, stable and independent structures of the finished carbon black. Highly structured carbon blacks have a high degree of linkage and branching, while in low structured carbon blacks this aggregation is weak. The aggregates then form loose agglomerates, the so-called secondary agglomerates, which are bonded by weak Van-der-Waals interactions. These carbon blacks are also called "fluffy". The void volumes between the aggregates and agglomerates, usually measured as absorbed volume of dibutyl phthalate (DBP) in a specified quantity of carbon black, describes the terminus "structure" of carbon black. The carbon black structure reflects the number of primary particles in an aggregate and their degree of branching. The specific surface area of an industrial carbon black is defined primarily by the particle geometry. A commonly used method of determining the specific surface area is the adsorption of cetyl-tri-methyl-ammonium-bromide (CTAB). CTAB adsorption is the closest method for determining the geometric surface area, i.e., the surface without pores, which correlates to the surface available for the rubber. Therefore, CTAB adsorption allows conclusions to be drawn regarding the application/technical behavior of the carbon blacks in a rubber compound. The specific surface area is also associated with the primary particle diameter, i.e., the smaller the particle diameter, the larger the specific surface area. The two quality parameters specific surface area (CTAB) and carbon black structure (DBP) are the most important variables for characterizing carbon blacks. On the left hand side of Fig. 5.2 an individual carbon black aggregate under a scanning electron microscope can be seen. The middle part shows high structure carbon black with a high degree of branching and the right hand part shows a low structure carbon black with a low degree of branching.

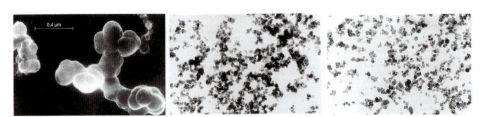

Figure 5.2 Photographs of carbon black particles with a scanning electron microscope. Source: Degussa AG

Fluffy carbon black, which consists of secondary agglomerates, is very difficult to handle. Therefore, fluffy carbon black is generally pelletized. However, industrial carbon blacks are also available as powder carbon blacks. Among pelletized carbon blacks a distinction is made between dry, wet, and oil pelletized carbon black. Pelletised carbon blacks provide considerable advantages compared to powder carbon blacks in regard to ecological, economical, and application-related issues, such as low dust loading, good flow behavior, good filling and metering capabilities, high bulk density, lower space requirements during transport and storage, reduced transport and storage costs, as well as faster incorporation and wetting [11, 12].

In the field of rubber processing mainly wet pelletized carbon blacks are offered [10]. During carbon black pelletizing, various pellet size grades evolve. The largest proportion is concentrated in the size grades 0.25–0.50 mm, 0.5–1.0 mm, and 1–2 mm, as can be seen in Fig. 5.3. Here, it becomes clear that the individual pellet hardness is dependent on the size of the pellet and the average pellet hardness. In addition, increasing pellet hardness leads to a larger distribution of the individual pellet hardnesses. The mean pellet hardness, which is normally stated in certificates, only provides partial information, namely an averaged value.

If we take a look at pelletized carbon blacks, it can be seen that pellet hardness plays a decisive role for the following mixing process. On the one hand, the pellets must be hard enough so that they are not too easily destroyed during transport or in the conveying processes. On the other hand, pellet hardness must be low enough so that the dispersion process of the carbon black is not influenced unfavorably. These two requirements are contrary to one another.

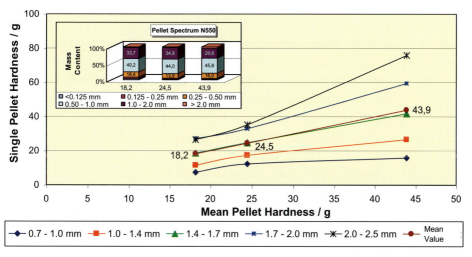

Figure 5.3 Single pellet hardness as a function of the mean pellet hardness and the particle size classes of N550

Ideally, during the mixing process the carbon black pellets are returned to their aggregate state. The secondary agglomerates play a decisive role here. The stronger the interaction between the aggregates, the higher the required forces to separate them again [13]. Boonstra and Medalia examined the dispersion process of carbon black in great detail [14 – 16]. For instance, they determined that large agglomerates exist after very short mixing times, which are destroyed bit by bit by the effects of shear forces in the kneader. From a formal point of view, the dispersion process can be split into four stages [17]: First the pellets are broken up and compacted. Next the polymer is incorporated into the aggregates. The polymer is first pressed into the voids in the carbon black agglomerates. Subsequently the polymer is incised into the carbon black. During this phase, the secondary agglomerates are broken up and phases of higher and lower carbon black concentrations evolve.

The void volume of the aggregates is then gradually filled by the polymer. In the next phase, the higher concentration phases are separated, although they are still coherent. In the last phase these coherent carbon black phases are distributed in the compound.

This process is shown schematically in Fig. 5.4. It is quite conceivable that the distribution rate is dependent, among other things, on the type of carbon black, the viscosity of the compound, and the compounding parameters. However, in practice the phases do not always run in the exact sequence stated above. Hence, it can be deduced that the parameter pellet hardness is an important quality parameter of carbon black. The significance is given by the fact that the task of the compounding process is to disperse and distribute all components of a rubber compound as well as possible. At the beginning it was already referred to the dependence of many compounds and end product properties on the degree of the achieved carbon black dispersion.

Figure 5.5 shows the mean pellet hardness and the maximum pellet hardness exemplarily for carbon black N550 recorded for approx. 70 deliveries to a rubber processing plant during one year in regard to the respective specification limits.

As shown in Fig. 5.5, there is a significant difference in values between the mean and maximum pellet hardness. It can also be determined that the maximum pellet hardness values are more dispersed than the mean pellet hardness. On the basis of this analysis it can be deduced that the influence of maximum pellet hardness should be considered higher than the mean pellet hardness in terms of an optimized quality assurance concept. It can also be presumed that the pelletizing process is very complex. The range between the specification limits is fully utilized. This means the pellet hardness shows the biggest allowable deviation. A variation

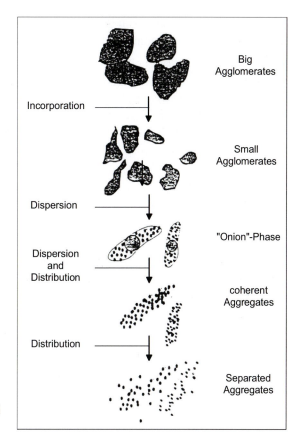

Figure 5.4 Model of carbon black dispersion procedure according to Shiga and Furuta [17]

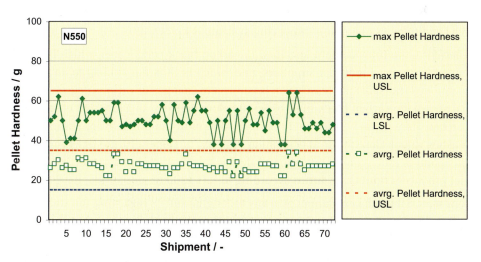

Figure 5.5 Mean pellet hardness and maximum pellet hardness of the carbon black N550 as a function of delivery [19]

of this magnitude has not been observed for most of the other carbon black quality parameters nor other raw material properties [18].

As described earlier, the pellet hardness setting causes problems in regard to the different requirements of the conveying, incorporation, and dispersion process. Pellets that are too hard can cause carbon black dispersion problems, especially in soft compounds. As discussed before, the degree of carbon black dispersion is linked to the quality of the compound and the products. Hence, the quality parameter pellet hardness will be examined here due to its significant, batch-related distribution. Another question can be derived, namely how can too soft carbon black pellets affect the process and product quality? If we analyze the previously introduced model of carbon black dispersion, it is conceivable that they could have a favorable effect on the process.

However, we are also faced with the question as to how the soft pellets behave during transport. In this case, transport includes both the delivery of the carbon black from the supplier to the rubber processor and the conveying process from the storage silo to the kneader. Depending on the quantity, carbon black is delivered in tankers, so-called Octabins, FIBCs[2], or sacks. If rubber processors use silos to accommodate the quantities they process, the carbon black is delivered in tankers in batches of approx. 20 t. If the carbon black is only a small component, sacked material is often sufficient. Octabins and FIBCs are used for small to medium throughputs.

It is obvious that during transport of carbon black, the pellets are subjected to undefined abrasion caused by relative movement. This produces the so-called carbon black fines[3]. These can increase considerably, if the carbon black is transported from Octabins or silos to the kneader with pneumatic conveying systems. The connection between the increase in the proportion of carbon black fines and the conditions in pneumatic conveying systems is known [20–23].

Therefore, the fines content specified in carbon black certificates, which is determined by the raw material suppliers, and which is thus not subject to any of the mentioned transport processes, can be subject to changes of various degree. Figure 5.6 shows the carbon black fines of N550 before delivery as a function of supplied batches. Here it can be seen that the fines content in the pre-delivery status rarely reaches values close to the specification limit of 7%. Generally, the fines content at this point is much lower than allowed by the specification limit. Hence, raw material producers fulfil their duties according to ISO 9000 or similar standards.

Raw material producers cannot give any other guarantees regarding the stability of carbon black pellets, as they have no control over the further processing in the pro-

---

[2] FIBC: Flexible Intermediate Bulk Container

[3] Definition of fines content: Carbon black particles < 125 nm, consisting of pellet fragments, dust, and very small pellets; determined according to ASTDM D 1508.

Figure 5.6 Carbon black fines of N550 before delivery as a function of supplied batches [19]

cessing plant. On the other hand, in [20] it was shown that the fines content can be much higher – up to 60%, depending on the conveying system. Therefore, the fines content is a quality parameter for carbon black, and under certain circumstances the state of this raw material is not accurately reflected in certificates as it enters the kneader. Consequently, the current quality assurance system may have a gap.

If the fines content is too high, this can cause problems in the mixing process. On the one hand, filling times in the mixer are prolonged by the increasing fines content, which makes the process less economically efficient. On the other hand, it is conceivable that the mixing process itself is affected [11, 12]. The latter can occur through compression of the fine material and/or filling level fluctuations in the form of dispersion problems, caused by the extended filling times in combination with suction effects (aspiration effects) [20].

A high proportion of carbon black fines also causes conveying instability during transport in pneumatic conveying systems. The fine material is prone to form large plugs, which can cause a temporary increase in pressure in the line. The pressure increases until the plug releases itself, which also causes the plug to accelerate. These effects can be recognized by the pressure fluctuations in the conveying system. Another possible problem in pipelines is baking of carbon black on the pipeline material. This causes the pipeline to get blocked after a period of time. The cross-section of the pipeline reduces and the air velocity increases. This increases all effects previously mentioned. In addition, pieces of the baked substance can break off the wall of the pipe and cause weighing errors and, depending on the state of compression, lead to undispersed carbon black in the rubber compound and contamination between different raw materials. High fines content also encourages separation by particle size in storage silos and intermediate containers. This can also cause plant interruptions. The parameters affecting the baking behavior of carbon black are specific surface area and structure, as well as the pipeline material that is used. At this point, it is unclear to which extent the individual variables contribute to this phenome. Hence, in addition to pellet hardness we also have to consider the fines content in terms of quality assurance in the rubber mixing room. Both parameters are also interrelated. Low pellet hardness favors high fines content and vice versa. Both quality parameters of carbon black should be investigated in a structured manner with regard to their effects on the mixing process and the product properties on the basis of the formulations and products in use. The respective variations of one quality parameter from raw material to end product will be tracked. However, in Section 5.2 we will first examine the changes in properties of raw materials in the rubber mixing room.

As explained earlier, the values of the remaining and not yet discussed quality parameters of carbon black, such as CTAB adsorption and DBP absorption, appear to be far less distributed. Obviously it is not a problem to produce carbon black with a constant quality in regard to these parameters. This is not the case with grit. This substance is basically a stone-like contamination in the carbon black, caused by the furnace lining. It is possible for small parts of the furnace lining to break off and contaminate the product. Grit is very hard and cannot be broken down by the shear forces in the mixing process. Consequently grit remains in the compound unless it is sieved out by a so-called strainer process. From a quality assurance point of view, unacceptable grit should be avoided because its detrimental effect is directly obvious, as defects of various sizes can be detected in the compound or product.

■ 5.2 Raw Material Changes in the Rubber Mixing Room

As already discussed, raw materials can change with regard to their quality parameters on their way from the supplier until their application in the kneader. Here, we will examine the effect of the increase in carbon blacks fines content. During the conveying of pelletized carbon black the fines content depends on the conveying conditions and the covered conveying distance. In principle, conveying systems are technically rough distinguished with regard to the air velocity. So-called dilute phase conveyors use high air velocities and low loads for the conveying of the goods and so-called dense phase conveyors use low conveying speeds and high loads.

5.2.1 Increase of Carbon Black Fines Content during the Conveying Process

In a circular conveying system, trials were carried out in order to determine the rate of decomposition of carbon black in dependence of the conveying parameters and the carbon black [11, 12].

Figure 5.7 shows the results of an increase in fines content in relation to the conveying cycles, the carbon black type, the piping material, and the conveying conditions. Four conveying cycles correspond to a conveying distance of approx. 270 m. In Fig. 5.7 it can be seen that with dilute phase conveying a much higher proportion of fines content sets in than with dense phase conveying. This effect is known and has already been described in earlier publications [20, 21], where the carbon blacks N330, N539 and N660 were analysed with respect to the conveying status and the conveying cycles. The reason for the higher fines content is the higher air velocities during the conveying process. Depending on the conveying method, the carbon black pellets are accelerated to 50 – 90 % of the velocity of the conveying air and, for example, at three times the speed of air they have nine times the kinetic energy due to the quadratic dependency. Particularly in elbows, this can destroy the carbon black pellets.

Figure 5.7 also shows for N234 and N650 that with the same conveying status, slightly higher fines contents are observed in the rubber pipe. This effect can be explained by the different coefficients of friction between the carbon black pellets and the different piping materials. However, what seems to be a disadvantage of the rubber pipe may be compensated by other advantages, as will be shown later.

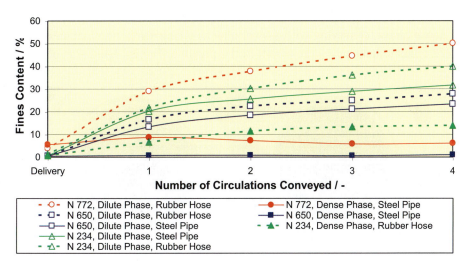

Figure 5.7 Increase of fines content as a function of the conveyed distance, the type of carbon black, the pipe material, and the conveying conditions

In the dilute phase conveying status in the rubber pipe it was observed that N772 experienced the most destruction and N650 the least. The difference between the fines contents of these two types was more than 20% after 4 cycles.

If we compare the dense phase with the dilute phase conveying system in the steel pipe, the fines content of N650 is approximately 10 times higher after the fourth cycle in the dilute phase conveying status. On the other hand, carbon black N772 behaves differently in the steel pipe in the dense phase conveying status. The fines content is already higher in the delivered condition; it increases after the first cycle and then seems to drop again. This curve suggests that the fines content bakes on to the piping.

Over all it can be seen in Fig. 5.7 that the increase in fines content is more obvious after the first conveying cycle (67 m). The increase of the curves becomes much flatter the longer the conveying distance becomes.

From [11, 12] it is further known that the higher the CTAB-absorption of the used carbon blacks, the higher the fines content during the pneumatic conveying. DBP absorption does not exhibit such such clear dependency.

5.2.2 Baking Behavior of Carbon Black in Conveying Pipes

The baking behavior of carbon black on the inner walls of conveying systems is another undesirable effect in addition to the increase of the fines content. As already assumed in the discussion of Fig. 5.7, the fines content of N772 baked on to the pipeline so that the carbon black fines content decreased during the further course of the conveying process. The test results are shown separately for different constellations of conveying condition vs. piping material.

In dense phase, conveying with a steel pipe as a conveying system, carbon black N772 tends to bake to a large extent in the pipeline. The left hand side of Fig. 5.8

Figure 5.8 Comparison of baking behavior after 5 (left) and 10 (right) circulations of N772 using dense phase conveying in a steel pipe

shows a picture of the inside of the pipe after 5 cycles and the right hand side shows the pipe after 10 cycles.

We see a clear increase in the layer thickness in the inner wall of the pipe from the 5th to the 10th cycle. After ten cycles the carbon black layer in some places is up to 20 mm thick.

In spite of the severe baking after 10 cycles shown in Fig. 5.7 it is still possible to convey the material in a stable manner because at this point there is no or very little fine material in the flowing air. But this also means that it would not be possible to determine the problem of carbon black pellet destruction in pneumatic conveying systems in this case on the basis of pressure fluctuations on the conveying system pressure gauge.

When a "higher structured" carbon black (N 650) was used, fewer baking effects (compared to N 772) were observed. The better conveying properties also caused a lower level of material distortion as the increase in the fines contents for each cycle was lower. It can thus be expected that higher structures lead to less "baking effects". Thus, very high active carbon blacks as for example N234, will have more favourable conveying properties in dense phase transportation.

In the dilute phase a similar effect can be observed. In a steel pipe conveying carbon black N650, considerably less baking was discovered and no major effect on the air mass flow was seen. Under the same conditions, carbon black N234 does not bake at all in the pipeline, as can be seen in Fig. 5.9.

As the carbon black is very light-absorbing, the inside of the pipe is illuminated with a torch in the left picture. At some points the light is reflected by the smooth metal surface of the pipe. In other words, no mentionable coating has formed. For N772, which was not tested in this constellation, we would expect the highest level of baking compared to N650 and N234.

Figure 5.9 Comparison of baking behavior after 5 (left) and 10 (right) circulations of N234 with the dilute phase conveying principle using a steel pipe

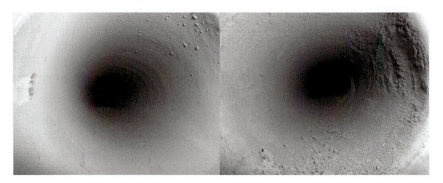

Figure 5.10 Comparison of baking behavior after 5 (left) and 10 (right) circulations of N650 with the dense phase conveying principle using a rubber hose lining

Carbon blacks N772, N650, and N234 show no noticeable baking in a pipe lined with a rubber hose. As an example, N650 is examined in more detail. Figure 5.10 shows the pipe after 5 (left) and 10 (right) circulations.

No large areas of baking can be detected in either picture. At several points smaller, crumb-like structures can be seen on the surface. However, due to the elastic rubber skin in the inside of the pipe, this type of baking only lasts a short time, as the elastic inner skin can cause a pumping effect. The difference between the inner diameter of the steel pipe and the external diameter of the rubber liner allows this movement. If the pressure in the pipe rises, the liner stretches and vice versa, if the pressure drops. These pulsating movements release any baking that may have occured. A pulsating release of areas of baking on the inner wall of the pipe has a kind of self-cleaning effect. This behavior was experienced in all experiments carried out with the rubber inner liner as a conveying system. This demonstrates the previously mentioned advantage of the rubber pipe justifying the rather higher fines content. Because of the lower fines content with dense phase conveying systems with a rubber inner liner, this effect is already negligible.

Using the dense phase conveying principle with a rubber inner liner, the condition of the dense phase conveyor was examined for carbon black N234. A comparison to dilute phase conveying with a rubber inner liner shows that the change in the conveying method had no influence on the baking behavior. The condition of the pipes is very similar to the one shown in Fig. 5.10.

5.3 Effect of Variations in Raw Material Quality Parameters on the Mixing Process

In the following the effect of long chain branching of EPDM polymer and the fines content as well as the pellet hardness of carbon black on the mixing process will be examined. In addition, two examples will be discussed explaining the effects of the delivery form of small chemicals. Also variations from the correct weight for major components were examined by changing the weight in the recipe by +/- 10%. Afterwards the impact of these "incorrect weighings" on the compound- and product characteristics were analyzed.

5.3.1 EPDM Long Chain Branching

EPDM long chain branching will be examined using an automotive seal profile (car side window). This formulation usually contains a highly branched polymer A that was replaced with the less branched types B and C. The other polymer quality parameters were practically identical.

Figure 5.11 shows the specific mixing energy during the first mixing phase – normally called mastication phase (NK) – as a function of differently branched polymers. As can be seen, the specific energy consumption for the highest branched polymer A is lower than the one of the less branched polymers C and D.

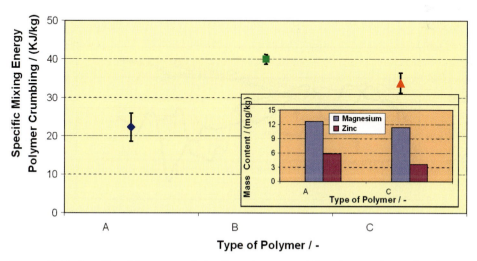

Figure 5.11 Specific mixing energy during the polymer crumbling phase and mass fraction of processing aids added to the polymer bale as a function of long chain branching; kneader: W&P GK 90E, rotor geometry PES3 [19]

This result seems to be in contradiction to the results mentioned in the literature [24, 25]. There it was determined that the specific crumbling energy rises as the degree of long chain branching increases. This can be observed with polymers B and C. The higher branched polymer B absorbs the most energy. Tokita, White, and Shih reported that highly branched polymers are drawn into the nip of a roller mill only with difficulty, not until a low degree of viscosity, favorable for integration, is achieved [26–29].

An explanation for the observations in this series of experiments can be found in the presence of processing agents contained in the bales of polymer A. A chemical analysis of A and C showed that the former contained slightly more magnesium and around 50% more zinc. The element zinc suggests the presence of the processing agent zinc stearate, a soap-like substance. Obviously this causes the polymer to break down easier under the influence of shear forces.

If all remaining ingredients of the formulation are added to the machine after the crumbling phase it can be seen that after the signal "ram down" the power consumption of the machine increases rapidly due to the increased filling level in the machine (see Fig. 5.12). The ingredients of the formulation are drawn into the mixing chamber by the rotors until the first maximum power level is reached. Here, this process takes approx. 20–25 seconds. Then the power input drops again, due to the wall slip effects until a minimum power level is achieved. In this phase, the mixing chamber contains a multi-phase system comprising a highly viscous polymer melt, dispersed bulk materials, liquid plasticizer, and various chemicals with different consistencies [4]. During this process phase the polymer perfuses fillers such as carbon black. The wall slip effect that leads to the reduced power input can be explained by the different coefficients of friction between the

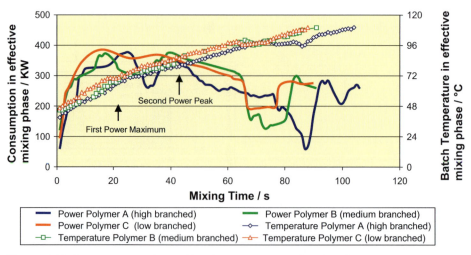

Figure 5.12 Power consumption and batch temperature during the effective mixing phase as a function of mixing time and the EPDM polymers' long chain branching [19]

mixed material and the mixing chamber walls and rotor surfaces: the fine bulk materials and the liquid plasticizer significantly reduce the coefficients of friction until a semi single-phase mixture has been created. The compound slips.

As the mixing process progresses, all ingredients of the formulation are incorporated into the compound and the power input increases again until the second maximum level is reached. Now a quasi single-phase mixture has formed (achievement of the BIT).

If the power curves are analyzed with respect to the polymers or the degree of long-chain branching, the following can be determined: the curve of the most linear polymer C suggests that it is the quickest to reach the ram end position, as the gradient is the highest. Carbon black incorporation (BIT) is also the fastest. It can also be seen that the minimum for this polymer in the wall slip phase is the least pronounced, while in the case of the highly branched polymer A it is most pronounced. This observation suggests a dependency on the long-chain branching, although slightly different start temperatures in the mixing process remain unconsidered. The effect can be explained as follows:

During the mixing process high levels of energy input are required to induce high shear- and elongational deformations. In order to incorporate other formulation ingredients, the surface of the polymer must be increased as much as possible. This requires the so-called laminar mixing process. Polymers with a high loss angle δ are conducive to laminar mixing. The different behaviors of polymers A and C, which differ most with respect to long-chain branching, can be seen in Fig. 5.13. It shows the loss angle δ of these polymers as a function of the polymers' testing temperature. In Fig. 5.13 it can be seen that the higher branched polymer A reaches the same loss angle δ at higher temperatures than the linear polymer C. As

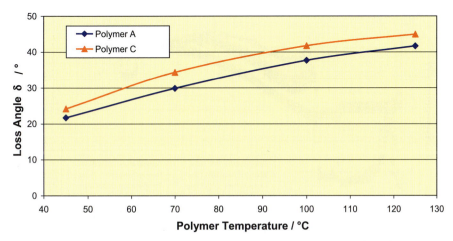

Figure 5.13 Loss angle δ of polymer A and C as a function of the polymers' testing temperature; testing frequency: 0.8 rad/s; amplitude: 0.5°; (RPA 2000, Alpha technologies) [19]

the temperature of the batch rises during the mixing process due to the dissipated energy, there is a direct connection between the described behavior of the polymer and the mixing process. As the level of long-chain branching increases in the polymer, the same value for the loss angle d is achieved later in the process, because of a lower degree of branching. This is why fillers are incorporated more slowly.

5.3.2 Carbon Black Fines Content

The fines content is typically provided by the carbon black supplier only in its condition before dispatch. But, subsequent transport processes are differing considerably at rubber processors, a fact that leads to different fines contents of carbon black in the condition before mixer charge. Here, it will again be referred to as the grain destruction independently of the applied conveying technique [11, 12, 20–23].

Figure 5.14 shows the power absorption and the temperature curve as a function of the mixing time and the carbon black fines content for an EPDM formulation (car door seal) in which the fines content was varied systematically.

It can be determined that the first power maximum occurs after all formulation ingredients have been added (except sulphur) due to the carbon black fines content. The effect can be explained by the wall slip phenomenon that was already described. As the carbon black fines content increases, the wall slip phase intensifies directly after the ram is set, which reduces the power maximum and subsequent minimum as a function of fines content. The theory is supported to a certain extent by the following consideration: the bulk density of carbon black

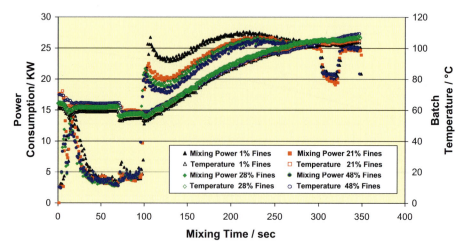

Figure 5.14 Power consumption and batch temperature as a function of mixing time and fines content; polymer A (highly branched) [19]

decreases with an increase in fines content. For example, un-destroyed N650 carbon black has a bulk density of 360 g/dm^3, while the same carbon black with 48 % fines content only has a bulk density of 295 g/dm^3. With varying bulk densities of the carbon black the filling level in the mixer at the start of the mixing process changes. In this example the filling level of the mixer would rise by 10 % due to the fines content if, instead of un-destroyed carbon black, a carbon black with 48 % fines content was used. Consequently, it would be expected for the power consumption at the first power maximum to rise, which is, however, not the case. Apparently, the fines content has a greater effect on the intensification of wall slip, which means a reduction in power consumption at this point in time.

During this test series, the highly branched polymer was also replaced with a less branched one. Here, a similar effect occurred as was already described earlier. However, it can be determined that the effect of fines content on the first power maximum also decreases with a decreasing degree of branching of the polymer. As discussed earlier, this can be explained by means of the higher incorporation speed of fillers.

5.3.3 Carbon Black Pellet Hardness

Another important raw material quality parameter to be discussed is the pellet hardness of carbon black. On the one hand, the pellets must be soft enough so that they can be dispersed and on the other hand, they must be hard enough to endure several transportation processes from the supplier to the mixer.

Figure 5.15 shows typical mixing curves for a compound for brake membranes. In these experiments, the pellet hardness was varied (20 g/3 g/45 g). Similar to the results for fines content, the power maximum is reduced for lowest pellet hardness (20 g) after addition of carbon black, stearic acid, and oil. Obviously, the size reduction of the softer carbon black pellets after addition into the kneader and after lowering of the ram increases fines content. However, the power maximum for the medium hard (30 g) and hard (45 g) carbon black pellets does not differ. In the subsequent wall sliding phase of the multiphase mixture in the kneader, a distinctive minimum is emerging for the soft carbon black pellets.

Up to this minimum, the power curves of the two pellet hardness settings cannot be differentiated. Only after the repeated power increase the hard carbon black pellets induce higher power consumption. This observation can be explained by the incorporation of bigger particles as it has to be assumed that the hard carbon black pellets are not yet broken up at the beginning of the mixing process. For EPDM, the Mooney Viscosity decreases nearly linear with increasing dispersion degree. After the addition of the remaining formulation contents, the power consumption practically does no longer differ for varying pellet hardness. However,

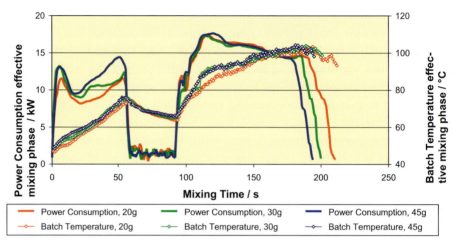

Figure 5.15 Power consumption and batch temperature depending on mixing time (effective mixing phase after addition of all formulation ingredients) and the pellet hardness of the examined carbon blacks [19]

the mixture containing the hard carbon black pellets is discharged first. In this last mixing phase, the batch temperature increases faster which can only be explained by the higher power consumption during the carbon black incorporation phase, although the batch temperature in the first mixing phase does practically not differ. This means that process changes caused by raw material cannot be predicted during the mixing process by means of temperature data analysis and therefore cannot be controlled.

As the mixing process was temperature controlled in this case, the faster temperature increase in mixing phase 2 causes the earlier discharge of the batch containing the hard carbon black pellets. This has a negative effect on batch quality [19]. If one tried to use the power absorbed by the kneader vs. the power consumption as a criterion for process control it would be logic that the speed would be reduced at higher power consumption. As a consequence there would be lower shear forces for the disaggregation of the harder carbon black pellets available, an effect that would have to be regarded as counterproductive. Therefore, general process- or even quality control is very difficult or even impossible without knowing the effect of raw material quality parameters.

It has to be noted that the higher power consumption during the incorporation time of the hard carbon black pellets is a characteristic of highly active carbon blacks. Semi-active carbon blacks will not exhibit the same behavior. However, the destruction of soft carbon black pellets obviously occurs independently from the activity level. An explanation for this phenomenon can be seen in the pellet hardness spectrum of the carbon blacks as a function of their activity. With increasing activity, the pelletizing process of carbon black becomes more complicated while the pellet hardness spectrum extends [11, 12].

5.4 Delivery Form of Sulphur

In the following, the effects of the delivery form of the small chemicals that are used in combination with kneaders with different technical standards will be presented. Although it is difficult to generalize, this example does illustrate the effects of raw material quality parameters in combination with the mixing technology on the quality of the end product. Rubber sleeves used to protect steel coils that are inserted on mandrels will be examined. These components generally have a very large volume and can consist of up to 150 kg of rubber compound. Consequently, there is considerable loss, if these components do not achieve the required quality.

The compounding process for the production of these sleeves was carried out in a GK 12 kneader built in 1963 with a gross volume of 38 liters. This kneader has a very special design that is only used in special industry branches today. The three-blade rotors are arranged diagonally above each other and the manual discharge door is located at the front end of the machine. The machine can only be operated at a speed of 34 rpm and the ram is activated pneumatically. In addition, the technical condition and wear of the machine can be described as extremely critical. The soluble sulphur, which is part of the curing system, is added on to the roller mill in the form of polymer-bonded sulphur pellets. Examining the type of process it would be expected to see an inadequate product quality. Therefore, the mixing process was first analyzed with the aid of the RELMA method [33] on the basis of compound samples taken from the rollers. And in fact, very high variation coefficients in regard to the sulphur were found in comparison to other investigations, which appear understandable in light of the process technology. If differing local sulphur concentrations are present and these also change from compound to compound, it is conceivable that this will influence the product quality. As shown in [19], the sulphur distribution often depends on the distribution of the carbon black. On the other hand, the product quality of the rubber sleeves that were investigated is known to be very good. This issue was investigated very simply and pragmatically within a series of tests. On the one hand, the sulphur dispersion was determined on the basis of samples that were taken along the process chain while on the other hand, the compound was also produced with a much more modern kneader, described as mixer 2 in the following. There is no need to provide a more accurate description of this mixer as it is insignificant for this study.

The results of the process analysis are shown in Fig. 5.16. The samples were taken after the milling process, after the extrusions of strips, and after production of the sleeves (after vulcanization). The average intensity ratio describes the concentration of sulphur in the mixture. The coefficient of variation allows a statement about the distribution of sulphur in the mixture, where high values stand for poor distributions. If the sulphur concentration in the mixture is identical within normal tolerances, the coefficients of variation can be directly compared.

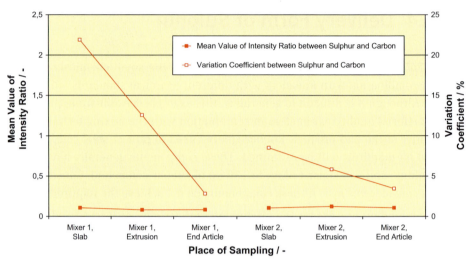

Figure 5.16 Process analysis of the manufacturing of big volumetric articles with the help of the RELMA unit as a function of the mixer applied

As can be seen in Fig. 5.16, the samples taken after milling exhibit high variation coefficients. But it can also be seen that the sulphur distribution along the process chain improved significantly. In particular the improvement in the condition of the extruded strips, which in this process are developed into sleeves via a coiling process to form the finished product, is astonishing as no further shear is introduced to the compound. The constructed sleeve is simply coiled and vulcanized in an autoclave. Analyzing the alternative mixing process in a more modern machine, where the sulphur has already been added to the kneader, we see a much better level of sulphur distribution after the mixing and milling process. Another improvement in the sulphur distribution also takes place along the process chain, although this is less pronounced. But it was also seen that there is no more significant difference in the sulphur distribution in the final product. In both cases sulphur distribution is rated as good. Variations in sulphur distribution cannot be the result of weighing differences, as the relative sulphur concentrations are constant in the compound. Hence, the sulphur distribution to be achieved in the finished product is independent of the condition of the mixing technology used.

As a result it can be interpreted that products characterized by high mass and wall thickness require long vulcanization times because of the generally low heat conductivity of rubber compounds. This product, for example, was vulcanized in an autoclave for four hours at 140 °C. Since soluble sulphur was used, its distribution can improve through diffusion during vulcanization. Diffusion processes are mainly dependent on temperature and time. Consequently, the distribution of the sulphur improved considerably during this process stage, although no shear was introduced to the product.

Another conclusion to be drawn from the investigation is that the entire process should be considered in order to classify the achievable product quality. Once again, the delivery form of the sulphur is responsible for the effects encounterd. On the basis of this example it can also be predicted that tests on compounds would almost certainly forecast poor product quality.

■ 5.5 Weighing Accuracy

Another important subject is the weighing of raw materials, which of course should always be as precise as possible.

Normally, the polymer and small chemical ingredients of formulations are weighed manually in the rubber processing industry. A double check is made when a computer-supported weighing system is installed in addition to the operator. For economic reasons, carbon black and plasticizer oils are usually weighed fully automatically. In practice it is not possible to weigh these ingredients completely accurately, as silos and weighing systems are connected by pipelines and the addition of raw materials is interrupted by the corresponding blocking devices just at the precise moment when the target weight is reached. A residual flow of the material – which is virtually impossible to estimate – remains in the pipeline between the cut-off device and the weighing system and continues to flow. Hence, it is only possible to comply with the target weight within certain tolerances. More precise weighing systems are available on the market, however only at higher investment costs. These systems often operate with a residual flow correction. Before they can be employed successfully, the effects of the formulation tolerances on the quality of the compound and the end product have to be determined. If tolerance violations are recorded during the compounding process, the minimum requirement is that the respective compound is examined in more detail if it is not in fact completely rejected. As carbon blacks and plasticizer oils are often significant components of the rubber compounds, they have a significant influence on the process behavior and the quality of the product.

The compound used in the investigation is a CSM compound applied as the covering layer of a low pressure power steering hose. The hose construction consists of a top layer, an inner layer, and a barrier layer made from polymer. During the manufacturing process the quantities of semi-active carbon black N550 and plasticizer oil were varied. During the subsequent manufacture of the hose the parameters of the other layers were not varied.

The two selected raw materials, carbon black and plasticizer oil were added in combinations of 10 % below the normal weight, the precise weight, and 10 % above the normal weight. In other words, a total of nine different compounds and hoses were produced. A two-phase mixing process was employed. Sample E is the reference compound, where the carbon black and oil quantities were not varied.

Within the scope of this investigation other tests were carried out on the compounds, which cannot be presented here in detail. These are tests that are normally carried out in the laboratory of a rubber-mixing room, such as Mooney Viscosity, rheometer, Shore hardness A, compression set, mechanical tests, and other product-specific tests. In addition, the dispersion quality of the fillers is determined according to the Phillips Scale. The results can be summarized as follows [31]: the Mooney viscosity, cross linking speed MH-ML, elongation at break, tear propagation resistance, and the modulus as well as Shore hardness A clearly show the variations in ingredients, while tensile strength does not show clear changes. Here, the distribution of the measured results is much larger than for the other parameters. The filler dispersion determined according to the Phillips Scale also does not map the variations. However, on the basis of elongation at break, tensile strength, and compression set it can also be seen that the minimum requirements are clearly exceeded, while in the case of compression set they were not reached. Any influences that the quantity variations have on resistance to ozone, low temperatures, and abrasion can be classified as insignificant. There were no significant changes in the above mentioned results after simulated ageing or after swelling tests with ASTM oil 1. Even then there were no inadmissible shortfalls in the specifications, although it was not possible to allocate all the results to the respective variations.

It can thus be concluded that the tests that were carried out on the compounds with standard equipment map the variations in ingredient input quite well in some instances. Consequently, it would be expected that weighing tolerances would have a significant effect on the product quality of the power steering hose. On the other hand, it was seen that in particular the mechanical properties of the compound were exceeded in every case despite the quite considerable variations in the formulation. In this case it would appear that the specification limits have been chosen so broad that there can practically be no rejects. But this also raises the question as to what sense the tests have in terms of quality assurance.

If the results of the tests in Fig. 5.17 are observed, no clear connections between the measured burst pressure and the weighing parameters can be seen. In addition, bursting pressure tests carried out with both untreated samples and samples that were subjected to special ambient conditions showed no obvious dependence on the varied parameters. It seems that there were large safety factors in place for the product, which show values of between 100% and 200% above the specification values. The bursting pressure is a decisive quality feature for these hoses. They have to fulfil their function after ageing and when subjected to low temperatures. Figure 5.17 shows the bursting pressures after low tewmperature application. The

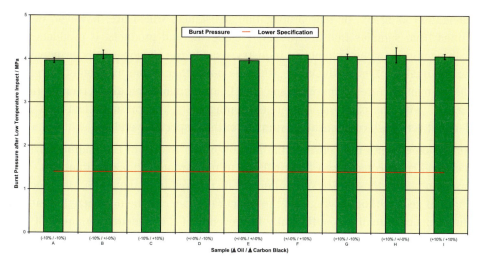

Figure 5.17 Bursting pressure of hoses after impact of low temperatures, as a function of the weighing tolerances of carbon black and oil applied; kneader: Francis Shaw (K6); rotor geometry: NR2

results show that all hoses exceed the lower specification limit by a factor of more than 2.5.

A similar result was obtained after ageing the hose for 16.8 hours at 140 °C and then another 152.2 hours at 120 °C. In this case, increased bursting pressure values compared to the non-aged condition were recorded. Thus, although the differences in the compound properties may seem significant there are virtually no deviations in the properties of the end product that could be attributed to parameter variations.

■ 5.6 Predicting Product Quality

Today, the quality of a compound in the rubber-mixing room is typically determined with the help of a Mooney instrument and with the so-called "Moving Die Rheometer" (MDR). In the technical rubber goods industry, as a rule all compounds are checked with these devices. If the Mooney viscosity and the respective key figures of the curing isotherms are within the specified tolerances, the compound is released and can be further processed into end products. Less frequently – in many cases just once per production order – other compound properties are determined and physical and/or mechanical tests are carried out on defined samples. End products are also only checked randomly regarding their required properties. Often these checks also involve mechanical properties such as tensile strength, elongation at break, or tear propagation resistance, which are, however, determined from samples that are prepared using the end products. If products are manufactured for the automotive industry, usually the vendor is fully responsible for the quality of its products, i.e., there is no further incoming goods control by the car manufacturer. If there are so-called field complaints requiring recall actions for cars with the corresponding re-equipping costs, the car manufacturers can demand very high compensation sums from the respective vendors.

As explained earlier, a considerable portion of the rejects in the manufacture of technical rubber goods can be attributed to inadequate compound qualities. The problem is that these rejects are often not discovered until the end of the process chain. It goes without saying that the reject rates can only be minimized, if either the compound quality is sufficiently improved or if the inadequate quality of the compound is detected as early as possible in the process chain. This means that inadequate raw material quality should be discovered before the beginning of the mixing process. If this is not possible, it should at least be possible to detect inadequate compound quality. Hence, in addition to the knowledge of adequate raw material specifications, it is important to install measures to determine the quality of compounds in order to correctly forecast the further behavior of the process and the quality of the end products. The performance of newer compound test methods such as RELMA, TOPO, and DIAS are described sufficiently in the literature [30, 32, 33]. In addition, the kneader will be examined as a possible instrument for quality assurance.

First, the results of the processes that were used will be listed in Table 5.1. The available process and quality information was collected as to the extent possible and provided in the form of a simple yes or no. Table 5.1 shows no clear correlation between product properties and process/testing parameters. There is no case in which the variation passes through to the end product and is properly predicted by all methods. It appears that the effects of changes in the respective raw material quality parameters are very low. The footnotes for the column "product properties" clearly illustrate that the influences on the quality of an end product has many facets. For example when different degrees of polymer branching were used for the car side windscreen seal this affected the process as well as the properties of the compound and the end product. However, this only applies when compounding is carried out in an intermeshing kneader. It should also be pointed out that the most important product property, compression set, which was determined for the end product, is independent of the parameter variation. Hence, the assessment "no" is conceivable for the effects of the raw material parameter on the quality of the end product.

To summarize it can be said that none of the methods used to test compounds was able to detect the variations in raw material quality parameters that were carried out within this study to a reliable extent. It would also appear that forecasting the

Table 5.1 Listing of the Results Describing the Different Compound Testing Methods in Case of Applied Raw Material Quality Parameter Variations [19]

Application	Parameter variation	Kneader	Effects on:						
			Mixing process	Mooney ML(1+4)	MDR	Mechanical properties	RELMA	TOPO, DIAS	End product properties
Car windscreen seal	Dd Polymer	GK90E	Yes	No	No	No	No	No	No
Car windscreen seal	Dd Polymer	GK160N	Yes	No	No	No	Yes	Yes	Yes [1]
Brake diaphragm	Dd Polymer	K7	Yes	No	No	No	Yes	–	No [2]
Dishwasher seal	Fines content N650	GK90E	No	No	Yes	Yes	Yes	No	–
Window seal	Fines content N650	K2	Yes	No	No	No	Yes	Yes	–
Tyre tread	Fines content N234	F70	Yes	No	No	–	Yes	No	Yes [3]
Dishwasher seal	Pellet hardness N550	GK90E	No	No	No	No	No	No	–
Building industry profile	Pellet hardness	GK90E	Yes	No	Yes	No	Yes	–	–
Inner layer air conditioning hose	Pellet hardness N550	K6	No	No	No	No	No	No	No [4]
Outer layer air conditioning hose	Pellet hardness N550	K6	Yes	No	No	No	No	Yes	Yes [5]
Brake diaphragm	Pellet hardness N220	BR1600	Yes	No	No	No	No	Yes	Yes
Brake diaphragm	Pellet hardness N220	GK5E	Yes	No	No	No	No	Yes	Yes
Air conditioning hose	Quantity of carbon black, oil	K6	No [7]	Yes	Yes	Yes	Yes	–	No [6]
Sleeve steel coil	Delivery form sulphur	GK12	No	–	–	–	Yes	–	No

1) Tensile strength and elongation at break vary, compression set does not.
2) The use of the MixCont system must be considered.
3) The influence of fines content is only shown for abrasion. No influence in all other measured properties.
4) Only the outer layer of the hose was examined, not the complete construction.
5) Only the inner layer of the hose was examined, not the complete construction.
6) The complete hose construction was examined although only the outer layer was varied.
7) These mixing curves could not be accurately evaluated due to very strongly fluctuating signals and imprecise data records.

quality of finished products is an even more difficult task. In this regard, the standard methods such as Mooney and MDR currently used in the mixing room as 100% control, seem to be particularly ill suited. On the one hand, the fluctuations in raw material quality parameters seem to be too small to be recognized by the test equipment, although the variations have proven effects on the finished products. On the other hand, several examples prove that other factors could be significant for the quality of a product. For instance, in the case of products with high safety factors it makes no sense to carry out highly sensitive tests in the mixing room. It also makes no sense to determine the sulphur dispersion in the compound when a soluble type is used, whose dispersion and distribution can improve significantly during vulcanization due to diffusion processes. In addition, the effects of raw material quality parameters also depend on the mixing technology that is used, as was shown in the case of pellet hardness.

It can be determined that it is not always possible to reliably forecast the product quality in the mixing room today with justifiable expenditure. Consequently, the quality assurance system in the manufacture of rubber articles can be described as lacking in certain respects. The quality of a finished product depends on the interplay between the raw materials, the production technology, and the product itself. But it has also been shown that forecasting the quality of a finished product can be improved substantially, if additional data from the mixing process, such as power and temperature curves, are consistently evaluated across the mixing time and more detailed test methods, as for example the RELMA method, are used.

■ 5.7 The Quality Assurance Concept "Future Mixing Room"

It is always desirable to control the mixing process. In the ideal case, all mixture components are put into the mixer and independently of raw material quality variations or other disruptive factors the result at the end of the process would be a constant compound quality. In addition to the better mixture quality, it could also be used to incorporate higher automation levels.

The following factors for the control of the mixing process, should be considered:

- Time
- Temperature
- Torque
- Energy
- Absolute number of revolutions, i.e., sum of the rotor revolutions with the ram at its lowered position

A time-controlled mixing process is the oldest way for reproducing the process. But as already shown earlier, variations, e.g., of the material parameters or the wall temperatures can lead to very different mixing results although the parameter time was not changed during the mixing process.

Measuring the compound temperature during the mixing process is one of the most difficult tasks of the sensor technology even today. Because of the very rough environmental conditions inside the mixing chamber, currently only very robust temperature sensors have an acceptable service life. They can cause, as shown in the example pellet hardness in Section 5.4, wrong temperature indications. The heat discharge to the cooled chamber wall leads to significant measuring errors. Because of their high mass the sensors are correspondingly inert and show reaction times of ca. 30 – 40 s (until reaching 95 % of the final value). In this time interval, the significant phase of the mixing process is possibly already finished. Final mixers used only for the second mixing step often exhibit total cycle times of 120 s. With regards to the stationary final temperature, today optimal sensors show tolerances below ± 5 °C. For sensitive controls (e. g., at chemical reaction controls) these errors are still too high. Despite these shortcomings, temperature control during the mixing sequence can be regarded as substantial progress compared to the time controlled operation method.

When the examples for the influence of the raw material quality parameters mentioned in Section 5.4 are put in the context of "mixing process control", it becomes clear once more how complex it would be to control the complete process. Especially the example of pellet hardness makes clear that the requests for a control strategy are very high and in many cases it is even impossible to have a "closed loop control" for compound quality based on process parameters.

A decisive obstacle to reach this target are contradictionary changes in material parameters, which could even compensate each other for example with respect to the power demand of the mixer (imagine a too high viscous polymer in combination with a too high fine content of carbon black). Although the power demand is not remarkably changed, big variations in end product properties or processability are possible.

There have always been positive approaches to improve individual phases of the mixing process using defined criteria, but no functioning overall strategy is known yet.

For example, in [34] a connection was established between specific energy [e_{spez}] and the viscosity degradation. The so-called Mix Cont System [35, 36] promises a better control of degradation of natural rubber [36]. This example and other tests [19] show that the specific energy is a useful criterion in the mastication phase to guarantee a reproducible mastication or blending phase of the polymer, respectively.

It becomes more difficult to use the specific energy input as a process control parameter if raw material parameters such as polymer branching degree, processing aids, pellet hardness, or fines content are also taken into consideration. It was explained in [19] that there is no constant correlation between the specific energy input into a mixture and the raw material parameters. Even the division of the performance chart into several mixing phases, such as the mastication- and the carbon black incorporation phase, does not always lead to the desired differentiation. This fact can be explained on the basis of the following example:

The specific energy input is a function of the mixing time (among other things). If, for example, the size of carbon black is severely reduced by the conveying process, causing a high fines content, it comes to an intensification of the wall sliding phase directly after the carbon black addition (see Section 5.4). Thus, the energy input in the mixture is reduced at the beginning of this phase and the incorporation time is extended (see Fig. 5.14). As a consequence, these two quite different incorporation phases (shorter and higher against less high but longer) lead to the same specific power input, but – of course – different carbon black dispersions. Consequently, a process control that kept the specific energy input constant would not work well in this case.

A temperature controlled mixing phase would also be problematic in this example. The wall sliding effects (in the case of a higher fine content) will cause a slower temperature increase and thus this mixing phase would be finished later. An extended mixing time, however, might not solve the dispersion problem as dispersion (see Chapter 2) is also effected by the absolute amount of the shear forces. So an extension at low power input might have no/not satisfactory effects.

So temperature controlled mixing phases – and phases which are related to the mixing time, such as the carbon black incorporation time – can be regarded as unsuitable for the appraisal of a constant process course, if the energy input is regarded as the defining criterion. In addition, the mixer-type is also important. Mixers with tangential rotor geometry show other feed-in behavior and different mixing characteristics compared to intermeshing geometries (see Chapters 1 and 2).

As shown in detail in Section 5.4 and in [19], the mixer reacts very sensitive to disturbance factors, e. g., resulting from raw materials. Deviations from the process data of the mixer correlate considerably better with the characteristics deviations of the final product properties than data of standard compound testing methods in the mill room (like Monney, MDR, etc.). This characteristic of the mixer can be used in order to produce a more constant compound quality.

The idea was to always run mixing phases such as the filler incorporation under same conditions (means not to "control" them, e. g., in terms of specific energy or temperature), and to analyze them. In case the performance chart differs significantly from a predetermined master curve, it is obvious that there is a deviation in raw material or during the feeding process. Although it is not possible to directly identify the cause, it is recommended to examine the respective compound in detail. On this basis, a quality assurance system was suggested in [19] which is displayed in Fig. 5.18.

The optimized quality assurance concept collects and analyzes all weight- and process data in a separate control system. If the data collected comply with the reference in a specified tolerance range, the compound can be released. With such a

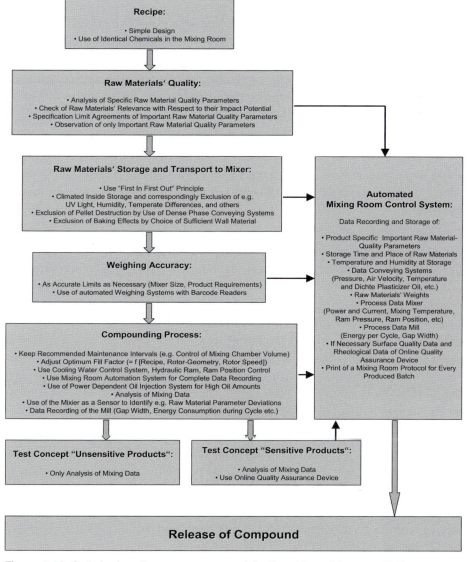

Figure 5.18 Optimized quality assurance concept for the rubber mixing room [19]

procedure in place, it is possible to predict the quality of the final product while the compound is still in the mixing room.

Nevertheless, in some cases it is useful to control several process phases. The simplest example for this is the temperature controller which uses the rotor speed as its control factor.

Considering the silanization reaction of a silica tread compound, which must take place at a constant compound temperature for a fixed time, the use of a temperature control device can be very helpful.

Another example is the application of the momentum dependent plasticizer injection for compounds having a high plasticizer loading. For dispersion reasons it is useful to add the plasticizer after the filler incorporation is finished. But in this case, there are no free carbon black surfaces available to absorb the oil. Therefore, it is possible that the mixer "falls asleep" which means the oil is taken into the mixture at a very slow rate. A lubrication film is forming between compound and mixing chamber walls or rotors, respectively, which leads to a power drop almost to the level of idle power consumption.

This problem can be faced with torque dependent plasticizer injection [37]. A control device analyzes the power input or the current consumption of the mixer, respectively and controls the oil injection valve. This prevents the "sleeping effect" of the mixer and the plasticizer is injected in portions into the mixing chamber.

Another issue to consider is the high dependence of the mixer on the fill factor. Especially for intermeshing mixers – which work on the principle of pressing the filler bit by bit into the mixer by means of the ram – it can be useful to control this process step. During this process phase the ram has not yet reached the final position, which means that the material still being under the ram does not take part in the mixing process at this point in time. With proceeding mixing time the fill factor increases with further moving down of the ram. Because the ram is subject to free forces (pressure from the rotor and material, ram pressure from "above"), the ram movement can adopt different characteristics in this process phase. Here, form and temperature of the polymer play a role as well as more or less accidental moistening processes with fillers etc.

A concept introduced in [38, 39] does not display an effective strategy to eliminate this problem. First, a compound is fed with constant ram pressure. When a satisfactory mixing process is achieved, the ram position/ram time characteristic is recorded as the nominal curve for process control.

A hydraulic ram offers the possibility to retrace this course during the succeeding mixing processes. While doing so the ram pressure is used as a set value. As shown in [40], this helps to obtain significant improvements with regard to the constancy of mixtures. For example, the deviations in Mooney viscosities of NR compounds that were mixed with a 45 l mixer with uncontrolled and controlled ram position could be reduced significantly. A noticeable improvement of the compound quality was recorded as well.

This control system also offers an interesting approach for fiber compounds. The problem here often is that when the fibers are compressed by the ram and then pressed into the mixing chamber, fiber nests are emerging that cannot be dissolved within the remaining mixing time. Here, the ram position control offers substantial advantages, because the ram position can be set according to any desired profile and so the compression of the fibers can be avoided.

The philosophy of a "ram position control" however is a contradiction to the strategy to keep the mixing phases unchanged and to analyze deviations for quality controls (see above). Controlling the ram postion will change ram pressure in some mixing phases. Otherwise, the advantages of a position controlled ram seem to be of such a high importance that in many case it seems to be advisable to abandon the possibility to use the mixer as a sensor in certain mixing phases and rather control ram position.

■ References for Section 5.1 to 5.7

[1] M. Bußmann: *KGK* **52** (1999), p. 742.

[2] E.-M. Arndt, M. Bußmann: "Gleiche Mischungsqualität von verschiedenen Produktionsstandorten – Qualitätssystem eines multinationalen Gummiverarbeiters" in "Mischen Extrudieren Spritzgießen"; VDI-Gesellschaft Kunststofftechnik. – Düsseldorf: VDI-Verlag, 2000.

[3] J. Hopf: "Einfluß der Eingangsqualität – Wie soll eine gute Rohstoffeingangskontrolle aufgebaut sein?" in "Mischungsherstellung – der Kernbereich der Elastomerverarbeitung"; VDI-Gesellschaft Kunststofftechnik. – Düsseldorf: VDI-Verlag, 1998.

[4] A. Limper, B. Barth, F. Grajewski: Technologie der Kautschukverarbeitung, Carl Hanser Verlag, München/Wien, 1989.

[5] W. Hofmann; Rubber Technology Handbook; Hanser Verlag, 1996.

[6] J. W. M. Noordermeer: Ethylene-Propylene-Diene Rubber in: Kirk-Othmer Encyclopedia of Chemical Technology – Fourth Ed., Volume No. 8 (1993) John Wiley & Sons, Inc.

[7] J. W. M. Noordermeer, M.J.M. Wilms; Processability of EPDM Rubbers in Internal Mixers – Dependence of Molecular Structure: *Kautschuk, Gummi, Kunststoffe*, **6** (1988), p. 558 – 563.

[8] H. J. H. Beelen; High Performance EPDM Polymers based on a New Technology of Controlled Long Chain Branching: Technical Information Bulletin DSM, March 1998.

[9]   H.J. Booij; Long Chain Branching and Viscoelasticity of Ethylene-Propylene-Diene Elastomers: *Kautschuk, Gummi, Kunststoffe,* **2** (1991), p. 128–130.

[10] G. Kühner: "Was ist Ruß?" Brochure Degussa-Hüls AG, Hürth 1996.

[11] H. Keuter, A. Limper, A. Wehmeier, T. Riedemann, K.H. Freitag: Feinanteilanstieg und Anhaftverhalten bei der pneumatischen Förderung von Ruß. *Schüttgut* **6** (2000) Nr. 4, p. 385–394.

[12] H. Keuter, A. Limper, A. Wehmeier, T. Riedemann, K.H. Freitag: Increase in fines content and adhesion behaviour in the pneumatic conveying of carbon black. *Rubber World,* Volume 224, No. **4/5** (2001), p. 29.

[13] G. Kraus: *Rubber Chemistry and Technology* **38** (1965), p. 1070.

[14] B.B. Boonstra, A.J. Medalia: *Rubber Age* **92** (1963), p. 892.

[15] B.B. Boonstra, A.J. Medalia: *Rubber Age* **93** (1963), p. 82.

[16] B.B. Boonstra, A.J. Medalia: *Rubber Chemistry and Technology,* **36** (1963) p. 115.

[17] S. Shiga, M. Furuta: *Rubber Chemistry and Technology* **58** (1985), 1/22.

[18] H. Keuter: Paper presented at internal 30 Montas-Meeting of the Mini Derucom Project (BRITE EURAM III Programm) on 2. April 2001.

[19] A. Limper, H. Keuter: Quality Assurance in the Rubber Mixing Room. Institut für Kunststofftechnik, Universität Paderborn, Gupta Verlag 2003.

[20] K.H. Freitag: Wirtschaftliche Aspekte bei der pneumatischen Förderung von Ruß; *Schüttgut* **1** (1995); p.115–117.

[21] D. Heep: Application and Optimization of a Dense-phase Carbon Black Conveying System in a Tire Works; *bulk solids handling,* Vol. **15** (1995) No. 1, p. 53–63.

[22] G. Thompson: Ultra low fines – a new approach; *Tire Technology International* (1996).

[23] W. Reiff: Improving Raw Materials Handling; *Tire Technology International,* (1996).

[24] J.W.M. Noordermeer, M.J.M. Wilms; Processability of EPDM Rubbers in Internal Mixers – Dependence of Molecular Structure: *Kautschuk, Gummi, Kunststoffe,* **6** (1988), p. 558–563.

[25] H.J.H. Beelen; High Performance EPDM Polymers based on a New Technology of Controlled Long Chain Branching: Technical Information Bulletin DSM, March 1998.

[26] N. Tokita, J.L. White; *Journal of Applied Polymer Science,* **10** (1966) p. 1011.

[27] J.L. White, N. Tokita; *Journal of Applied Polymer Science,* **12** (1968) p. 1589.

[28] J.L. White; *Rubber Chemistry and Technology,* **50** (1977), p. 163.

[29] C.K. Shih; *Trans. Soc. Rheology,* **15** (1971), p. 759.

[30] H. Keuter, D. Ackfeld, A. Limper: RELMA – ein Instrument zur Qualitätssicherung im Mischsaal; *KGK* 53 (2000) Nr. **10**, p. 566–573.

[31] K.-U. Kelting: Untersuchung von Rohstoffeinflüssen auf Mischprozeß und Endprodukt-Eigenschaften elastomerer Bauteile; unveröffentlichte Studienarbeit am Institut für Kunststofftechnik, Universität Paderborn; 2000.

[32]   A. Wehmeier: Entwicklung eines Verfahrens zur Charakterisierung der Füllstoffdispersion in Gummimischungen mittels einer Oberflächentopographie; Thesis Fachhochschule Münster, Dept. Steinfurth, 1998.

[33]   H. Geisler: Einfluß der Mischungsherstellung auf die Produkteigenschaften; in: VDI-K Seminar "Mischungsherstellung für Elastomere – Workshop mit praktischen Versuchen", 1999.

[34]   J. Teuber: Aufbereiten von Kautschuk und Kautschuk-Grundmischungen. Thesis FH Heilbronn, 1986.

[35]   D. Shaw: Mix Controller optimises mix cycle in real time; *European Rubber Journal* Vol. **183**, No. 5, May 2001.

[36]   M. Sarbatova, U. Petterson, S. Brassas: Advanced real time control system for rubber mixers. Paper presented at the International Rubber Conference in Birmingham, UK, 2001 on June 12 – 13.

[37]   W. Häder, H.-M. Monyer: Integrated control. *Tire Technology International,* June 1999.

[38]   J. Sunder: Regelung und Optimierung des Mischprozesses von Elastomercompounds im Innenmischer. Dissertation RWTH Aachen 1993.

[39]   J. Sunder: Möglichkeiten zur Verbesserung des Einzugsverhaltens am Innenmischer. Fachbeiratsgruppe Kautschuktechnologie des IKV 1988.

[40]   J. Sunder: Neue Regelungsmöglichkeiten am Innenmischer. Vortrag auf dem 16. IKV Kolloquium Aachen 1992.

■ 5.8 Rubber Compounding and its Impact on Product Properties

Typically it is assumed that the quality of the rubber compound depends only on the compound recipe or on the qualities of the raw materials used, while the influence of the processing steps is often underestimated. Nevertheless, with the growing demand on manufactures of elastomer parts to increase throughputs and at the same time to reduce costs, the mixing process gains more importance. For the mixing process, higher throughputs lead to smaller tolerances within which the required product properties have to be reached. Therefore, compound testing becomes more important in the production lines of batch processes. In today's practice, variations in compound properties are insufficiently identified by the analysis and characterization methods usually used. This leads to the fact that problems during mixing are often recognized after further processing steps or sometimes even after the final check of part properties. Both the processability of the compound and the required properties of the product should be judged by suitable compound testing methods.

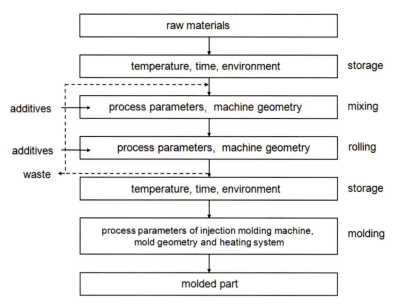

Figure 5.19 Influencing factors on part quality

When producing elastomer parts, the manufacturers process many different compounds that are manufactured using particularly adapted recipes. The mixing process turns out to be very complex, in particular because of the different states of material aggregation. Before an elastomer part can be produced, a rubber compound goes through several steps, which can affect the part properties significantly (Fig. 5.19). Compounding is typically achieved in the internal batch mixers.

Today, important aspects of the process cycle, such as optimizing machine, tool, or screw construction can be simulated, configured, and calculated before a part is developed. This is possible only if for each manufacturing step sufficient information is available about processes, boundary conditions, and the properties of the material to be processed.

The process relevant properties are:

- Flow/rheological characteristics
- Curing characteristics
- Thermal material behavior

These properties are affected mainly by the following factors:

- Raw material properties
- Compound recipe
- Conditions of storage and its duration

- Processing steps (internal mixer, rolling mill)
  - Duration
  - Number of mixing phases/stages
  - Mixing order

The causes for batch-to-batch fluctuations are usually contributed to the discontinuous processing methods, changes in thermal boundary conditions, and to varying coolant temperatures during batch processing. Variations in processing time, partial or semi-manual machine conditions, and the human factor all strongly affect compound and part properties. An overview of important influences on compound properties is shown in Fig. 5.20.

In the 1980s, a number of test and analysis methods for the characterization of compound and part properties were developed and constantly improved [1, 2, 3, 4, 5]. Today, processors still struggle with identifying the best method to predict processability and part properties. A solution is possible only if both compound mixing and downstream process are considered and sufficiently measured or monitored [6].

In 1952 Dannenberg [7] used the example of SBR to show that the mechanical properties of carbon black-filled cured rubber compounds strongly depended on mixing time. Ten years later, Boonstra and Medalia [8] followed up in more details. It was noted that the viscosity increased with increasing mixing time, i.e., with increasing dispersion degree and decreasing wear properties.

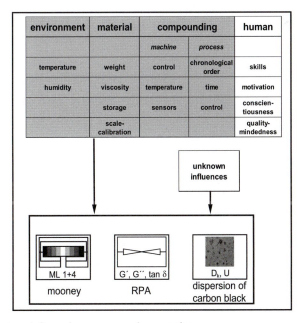

Figure 5.20 Factors influencing compound properties

Payne [9] showed that by using the same mixing process the amplitude of the shear modulus appeared to depend on the mixing time and/or dispersion. The dependence of physical properties on the number of rotor revolutions of the internal mixer and on the kind and structure of the used filler was also examined in detail. In particular, Hess [10, 11] examined the influence of process parameters (time, number of revolutions) as well as the effects of different surfaces and structures of carbon black in both SBR and EPDM compounds. As particle sizes served in particular dispersion, die swelling and tensile strength were investigated. The influence of process parameters and compound ingredients on the product properties were characterized by means of multiple linear regressions. The results of these and other work by Hess showed the importance of filler dispersion as a major/main cause of changes in compound and part properties. Investigations during a European Union project ("Minimum default in Rubber Compounding" [12]) showed the importance of changes in fine carbon black structure, humidity of the filler, and changes in filler properties.

With the exception of the methods for carbon black dispersion, all standard testing methods used for compound characterization mainly provide an evaluation of the processing behaviors. The focus here is the determination of the flow behavior and/or the vulcanization characteristics. The most important rheological testing methods used are the capillary rheometer and the Mooney-viscometer [13]. In industrial practice, the capillary rheometer has not found wide spread application due to its high test costs and problems caused by wall slip behavior. The crosslinking behavior is evaluated in practice using vulcameters [5, 14]. Often, the importance of the result of these testing methods is reduced because they do not really represent the process conditions. Therefore, testing methods have been developed that are better adapted to the processing steps. There are the rheovulcameter covering injection molding [15], or the visco-elastometer for the extrusion process [1], to mention just a few. In addition, the determination of elastic storage and loss modulus is possible using rheometers working with oscillating deformation [5].

■ 5.9 Testing Methods for Rubber Compounds

Compound control is used to describe the processability of compounds in the further processing steps. Depending on the desired compound or product properties and the required testing accuracy, there are a lot of methods currently available. The best-known methods are the Mooney viscosity and the vulcameter [16, 17]. These relative testing procedures are often used to optimize and analyze the production processes and to determine the relevant correlations between compound and part properties. Although a large number of relative methods is available, establishing meaningful measured variables is still problematic for elastomer processors. The fundamental question is which of the measured compound properties correlate better to processability and/or part properties. Further testing methods (absolute testing procedures) are used to accurately determine rheological material properties, such as viscosity and flow curves or rheological coefficients for the simulation of processing processes.

Flow characteristics are tested by

- Mooney viscometer (MV),
- rubber process analyzer (RPA)

Cross-linking/curing characteristics are tested by

- vulcameter (MDR),
- rheovulcameter,
- rubber process analyzer (RPA)

Filler and additive distribution are measured by

- carbon black dispersion,
- remote laser material analysis RELMA [deru2002].

In the following sections, a number of relative methods for compound characterization are presented.

5.9.1 Mooney Viscometer

The most established procedure is the measurement of the Mooney viscosity. The main operational area of the testing procedures involved is not process design, but quality control. This plays an important role, especially in rubber processing due to the discontinuous nature of mixing in the internal mixer [18].

The Mooney viscosity is determined according to DIN 53,525 [13]. The sample is sheared between the chamber and rotor surface by rotation of a flat disk with a constant number of revolutions ($n = 2$ min^{-1}). The torque is measured in Mooney units (1 ME = 0.083 Nm, large rotor). For the description of the elastic behavior of the sample, a relaxation parameter can be determined in addition to the standardized method. The rotor is then stopped and the value of the torque response after a certain time T3, is determined as a further rheologic characteristic – Mooney decay behavior (see Fig. 5.21).

With this measurement, an additional value for the shear viscosity function can be calculated using a Fourier transformation. The consistency factor K and the flow exponent n of a rubber compound can also be measured [19].

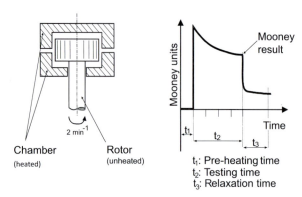

Figure 5.21 Test chamber and Mooney curve

The determination of the Mooney viscosity is widely used in the rubber industry because it is considered a simple and practical testing method for testing. It is used particularly for raw material tests (raw rubber) and for compound testing. It supplies a characteristic, device-specific viscosity value.

5.9.2 Vulcameter

The vulcameter is used successfully today in industry for the characterization of material behavior regarding the curing reaction [20]. Currently, using a rotor-less rotation vulcameter is preferred. The course of the maximum torque is measured, which in turn can be translated into a response of the material to a sinusoidal reciprocating shear deformation. Various information can be derived from the vulcameter curve (Fig. 5.22): the course of the vulcameter curve indicates the optimal setting of the vulcanization process (e.g., heating time control, using a heating time computer [Kamm98] based on the t_{90}-time). Here, the parameter t_{90} describes the time, at which 90% of the torque difference $M_{d, max} - M_{d, min}$ is reached. The times T_{s1} and T_{s2} describe the incubation period (beginning of the curing). All these parameters are used to establish suitable compound control [21]. In addition,

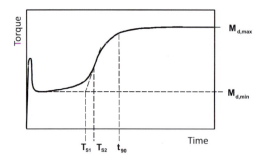

Figure 5.22 Selected characteristic values of the vulcameter curve

the turning point of the vulcameter curve (the maximum curing speed) is also used as a further characteristic value.

Additional normal force measurements with modern vulcameters (e.g., elastograph "Vario" [14]) are particularly interesting for expanding material systems (e.g., foam/sponge rubber). The testing data usually provide further information about the curing mechanisms, particularly during further processing (1st and 2nd derivative), if multi-level curing reactions are employed [14].

5.9.3 Rubber Process Analyzer

Alpha Technologies' rubber process analyzer 2000 (RPA2000) offers a number of possibilities for the investigation of elastomers [22]. This method is characterized by a very high testing accuracy; it is able to describe compound fluctuations with high sensitivity. The most important ranges of application of this equipment are represented in Fig. 5.23.

An uncured sample is inserted into the chamber. It can be tested before, during, and after the curing process. In addition, the sample is placed in the cone-shaped chamber, Fig. 5.24, which is similar to the conventional vulcameter. There are radial ribs for better moment transmission.

The test specimen undergoes sinusoidal reciprocating shear deformation. Frequency and amplitude (angle of strain) are selectable. The RPA 2000 measures the course of the resulting torque and calculates shear stress, the complex shear

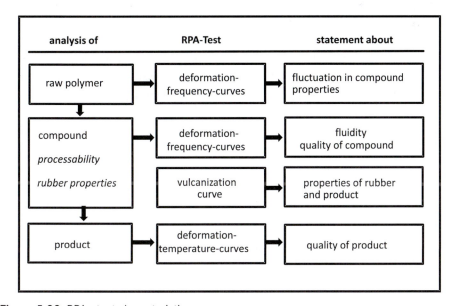

Figure 5.23 RPA - test characteristic

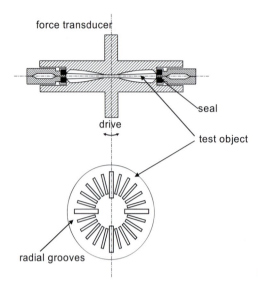

Figure 5.24 RPA 2000-test chamber [22]

modulus, and the complex viscosity as well as the elastic and viscous parts of the measured values.

The changeable parameters and their ranges are:

- Frequency: 0.002 to 33.33 cycles per second
- Angle (Strain): 0.05 to 90°
- Temperature: 40 to 230 °C

The resulting shear rate cannot exceed 30 s^{-1}. Due to the cone geometry of the testing equipment, the shear rate is generally constant within the sample and

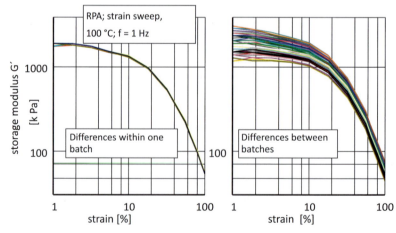

Figure 5.25 Dispersion comparison with the RPA data (strain sweep, 100 °C, f = 1 cycles per second, SBR standard injection molding compound, laboratory mixer GK-5 E)

linked to angle and frequency. The high reproducibility of the measurement and the high resolution regarding the differences between batches is shown in Fig. 5.25 for an SBR injection molding compound.

The left diagram shows the result of an RPA measurement for one batch. In order to test the reproducability of the method, 9 different samples taken from one batch were measured. The right diagram depicts testing of several batches of a compound, which were manufactured with different mixer settings. The width of scattering of the measurements confirms a high resolution of measurement procedure, as indicated in the left picture.

5.9.4 Carbon Black Dispersion Measurement

The distribution of carbon black particles and the size of the carbon black agglomerates in a rubber compound have an important influence on part properties. During the mixing process the carbon black agglomerates are broken by dispersive mixing and then incorporated into the polymer matrix. The higher the degree of particle destruction, the higher the specific carbon black surface becomes. This determines the effect of the carbon black as an active filler and/or reinforcement. A larger specific surface, which corresponds to a high surface to volume rate, leads to better part properties. The following properties are affected by carbon black dispersion:

- abrasion resistance,
- tensile strength,
- Young's modulus.

Both light and transmissions electron microscopy are suitable methods for the determination of carbon black dispersion. Another important method to analyze filler and/or chemical dispersion is the RELMA method [12]. More specific information can be found in the literature list at the end of this chapter.

■ 5.10 Factors Influencing Rubber Part Properties

For the production of elastomer parts all steps of the process chain must be considered, starting from compound mixing (inclusive mills or strainer), through the intermediate steps (e.g., manufacturing of stripes or plates), to injection molding or extrusion with a consistent quality assurance concept (Fig. 5.26). The relevant information have to be combined with process data as well as the properties of the

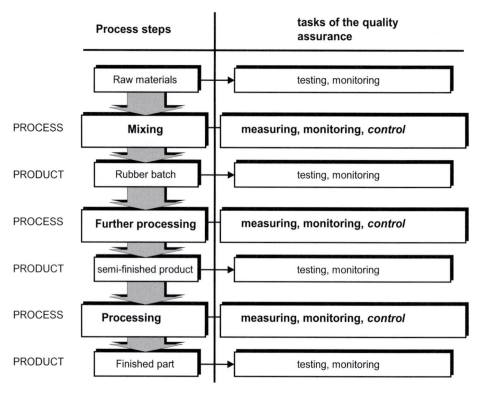

Figure 5.26 Production chain for rubber processing and tasks of quality assurance [23]

respective products (compound, semi-finished material, or finished part) in a QA concept.

On the one hand, the *product* properties are examined and compared to the specifications, on the other hand, the process parameters have to be measured during a *process*. With sufficient knowledge of the processes, optional process control is conceivable (shown in italics in Fig. 5.8). Each of these individual process steps affects part quality. Therefore, an analysis of the correlations between the process parameters of the respective manufacturing steps and the product properties is required.

In the following, we will describe the most important influencing factors of the mixing process on part properties (see Fig. 5.27). These factors are divided into the following groups:

- Mixing process → Compound properties
- Compound properties → Processing
- Mixing process, compound properties → Part properties.

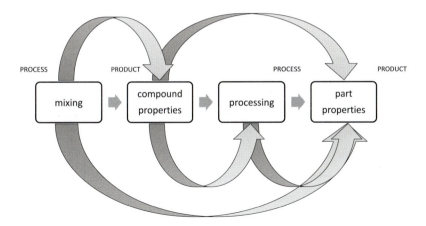

Figure 5.27 Influencing factors: mixing process - compound - product [23]

5.11 The Mixing Process

5.11.1 The Mixing Process and its Tasks

The mixing processes can be divided in two main phases:
- Production of the masterbatch(also multi-level)
- Production of the final compound.

These two phases can be realized within one process (single-step mixing process), or within processes divided into two or more steps, separated temporally and spatially from each other (Fig. 5.28). Figure 5.29 contains the typical process parameters.

In general terms, typical mixing procedures contain the phases of mastication, filler and softener incorporation.

Mastication
The mastication phase usually begins immediately after the rubber is being fed into the internal mixer (Fig. 5.29, A). During the mixing process in the mixer, new surfaces are constantly being created, so that a better penetration of the material layers is achieved due to the high shearing stresses. A simultaneous viscosity reduction fosters good distribution of compound components [Mic90].

Filler Incorporation
In this phase, fillers are added to the plasticized polymer (D). The ram moves down and mixing continues, leading to a 1st peak of the electrical power (E) and the first

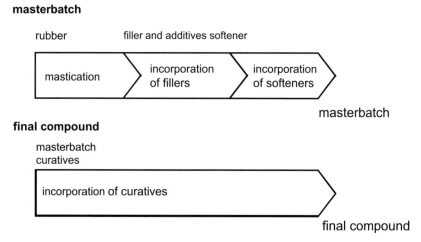

Figure 5.28 Process descripion in 2 – (n = 1) and multi-stage mixing process

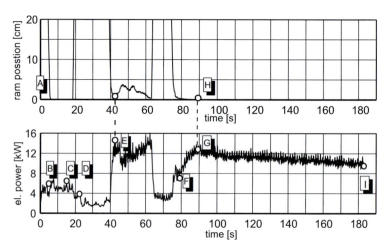

Figure 5.29 Power and ram position curve for a mixing process in a laboratory mixer (GK-1,5 E; Harburg Freudenberger)

minimum of the ram position curve. This position is reached due to the kinetic energy of the ram. During the incorporation of the fillers the ram moves towards its final position. However, before this position is reached, liquid plasticizers are added. The addition takes place at a time, when a small portion of the filler is still in the chute (under the ram). Thus it can absorb a portion of the plasticizer and facilitate oil incorporation. Moreover, the oil (plasticizer) addition becomes more reproducible (see Chapter 2).

Oil Incorporation
In this phase, the liquid additives are added to the rubber compound. At the beginning, the power curve decreases (F), caused by oil moistening the mixer housing and the rotor surfaces. After the free carbon black agglomerates have absorbed the oil and the oil is incorporated into the polymer matrix, the power curve increases. The ram reaches its final end position (H), simultaneously with the power maximum (G). Ejection (I) of the compound from the mixer takes place when the compound reaches a preselected temperature or a preselected time condition while the ram is down.

Final Mix
During final mixing, the cold masterbatch as well as the curing system are mixed in the internal mixer (or on the mill). In order to prevent a curing reaction of the compound in the internal mixer, the mass temperature in particular must be monitored and supervised. The completion of the process takes place before reaching the maximally permissible compound temperature.

5.11.2 Further Processes

Typical additional steps following the mixing process in the internal mixer are cooling and homogenization of the compound. These steps have the additional task to form the compound into different shapes for the final manufacturing process, e.g., extrusion, compression, or injection molding. All these steps are carried out with an uncured rubber compound. Table 5.3 lists some of these steps and their tasks in the production chain. Detailed descriptions of these processes can be found in [24, 25].

Table 5.3 Processing Steps and their Tasks

Step	Task
Mills	Cooling, homogenization
Strainer	Homogenization, clean
Refiner	Homogenization, clean
Batch off	Cooling
Calander	Assembling
Manufacturing of prepregs	Assembling

5.12 Factors Influencing the Mixing Process

During compounding the filler particles are dispersed and distributed by the mixing forces. The quality and intensity of the dispersion and the distribution of white and black fillers plays a particular role regarding compound quality and part properties.

The main goal of the mixing process is to break up and distribute the carbon black particles homogeneously by increasing the number of revolutions (Fig. 5.30), shear, or compounding time, as shown in Fig. 5.31.

After a certain mixing time a constant dispersion level can be achieved, above which no further improvement is possible. In addition, Fig. 5.31 shows that the carbon black dispersion is clearly dependent on ram pressure. Increasing ram pressure (increase in pressure in the mixing chamber) leads to an acceleration of

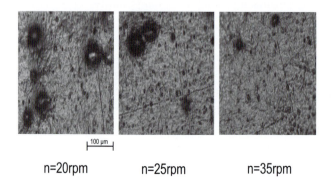

Figure 5.30 Dependence of filler particle size on the number of revolutions in the mixer

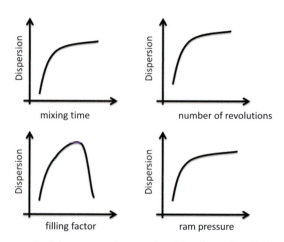

Figure 5.31 Influence of mixing parameters on the filler dispersion [26]

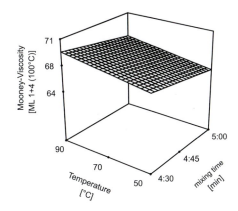

Figure 5.32 a Influence of mixer temperature and the compounding time on the Mooney viscosity [27]

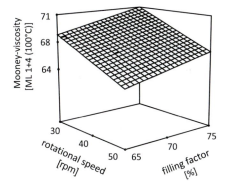

Figure 5.32 b Influence of number of revolutions and filling factor on Mooney viscosity [27]

filler incorporation and thus to an improvement in compound quality. The filling factor, which affects the pressure in the mixing chamber, cannot be increased above a certain value. If the mixer is overfilled, some compound components remain unmixed in the mixing chamber, forming so-called "dead spots". These effects are summarized in Fig. 5.13 and a detailed discussion can be found in Chapter 2.

Figures 5.32 a and b show the influence of the most important mixing process parameters on the Mooney viscosity. An increase in rotor speed and mixing time can lead to a lower viscosity, because of a better dispersion and incorporation of the carbon black, Fig. 5.32 a. Improvement in dispersion due to an increase in number of revolutions (within the range of 25 to 35 RPM) is shown in Fig. 5.32 b.

5.12.1 Influence of Plasticizer Addition on the Mixing Process and Compound Properties

Adding plasticizers into the rubber compound can have a crucial influence on the compound and later on the part properties. Figure 5.33 (top) shows an example of the consequences of incorrect selection of the point of oil addition, that is, an

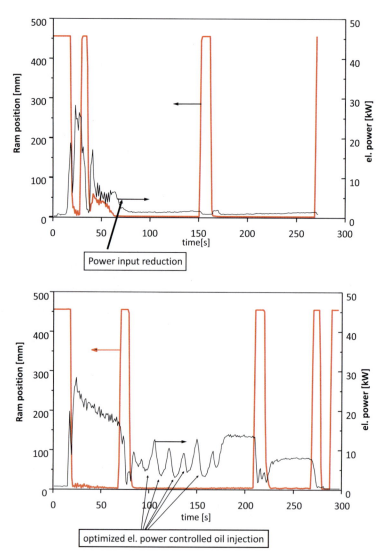

Figure 5.33 Mixing process curve for basic (top) and optimized (bottom) oil injection [27]

increase in mixing time and to poor dispersion. An improvement of the process cycle can be achieved by stepwise and/or torque-controlled oil addition (Fig. 5.33 bottom). It is also possible to add plasticizers simultaneously with the fillers.

Problems during oil addition often result in difficulties during subsequent treatments, such as the separation of the softener or "delamination" during the injection molding or extrusion process (elastomer layers separated by plasticizer layers). These processing difficulties often lead to part failures, e.g., surface defects (peeling, see Fig. 5.34) or flow and weld lines (Fig. 5.35), all of which strongly limit the functionality of the rubber part.

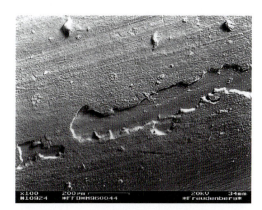

Figure 5.34 Surface defect - peeling problem

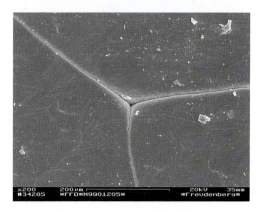

Figure 5.35 Flow defect - weld lines

5.12.2 Influences of the Mixing Process on the Injection Molding Process

An analysis of the total production process shows the importance of the compound properties on the further processing. Compounds with different shear modulus were analyzed on an injection molding machine. Figure 5.36 shows the heat flow between the plasticizing units as a function of the shear modulus of the compound. Because the injection molding process was carried out under constant machine settings, the cause for the observed changes (shown by the heat flow) can be exclusively attributed to the property differences between the rubber batches. In this case, a very high viscous compound (high modulus) leads to increased shear energy in the plasticizing/injection unit. This in turn leads to increased compound temperature, which can lead to increased thermal stress during injection of the compound and consequently to a reduction in scorch time and pre-curing effects.

As Fig. 5.37 shows, a compound with higher viscosity (x-axis) leads to an extension of the dosing phase, which corresponds to an increased pressure integral (y-axis). The measured changes in Mooney viscosity and the RPA data matched the changes in product properties. Their influence on shore hardness and the viscous portion of

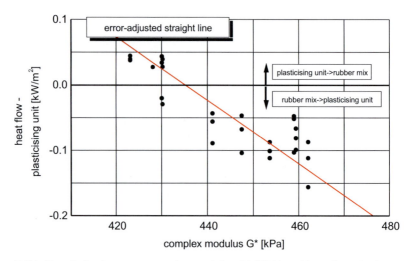

Figure 5.36 Correlation between complex modulus G* (RPA) and heat flows in the plasticizing/injection unit for an SBR standard compound [23]

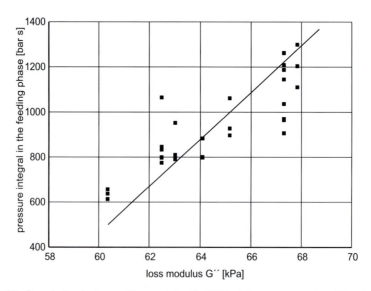

Figure 5.37 Correlation between the modulus G (RPA) of the compound and the duration of the feeding phase for injection molding [23]

the dynamic "feather/spring" stiffness (measured with the CAS Jidoka testing instrument) could be determined. Figure 5.38 shows that compounds with high Mooney viscosity lead to high dynamic spring rates of the rubber parts. Rubber compounds with high modulus result in rubber parts with high shore hardness, see Fig. 5.39.

It was shown that the viscous and/or storage modulus provide substantial information about the process and that the measured viscoelastic compound properties

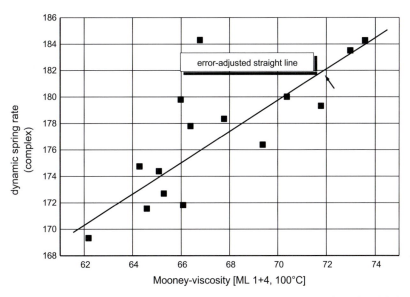

Figure 5.38 Correlation between compound properties and the properties of molded parts, Mooney-viscosity vs. dynamic spring rate for an SBR-standard mix [23]

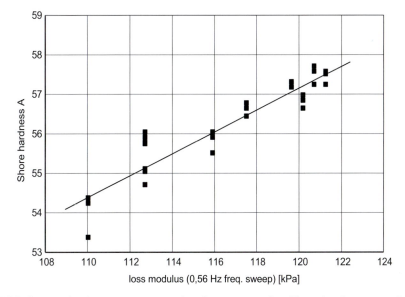

Figure 5.39 Connection between compound and part properties; Shore hardness complete modulus (RPA) for an SBR standard compound [23]

are well suited for the evaluation of compound processability and for achieving the required part properties.

Figure 5.40 shows the dependence of the occurence of flow marks on Mooney viscosity with different injection pressures. The occurence of flow marks was rated

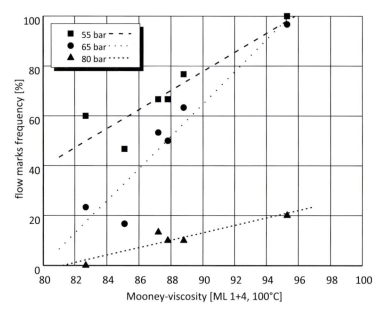

Figure 5.40 Surface defects as a function of Mooney viscosity

(100%: all parts show strong flow marks, 0%: all parts are free of flow marks). The results show the influence of compound properties and processing on part quality. A reduction in part surface defects can be achieved with lower compound viscosities and higher injection pressures.

5.12.3 Influence of the Mixing Process on Extrusion

The dispersion and distribution of fillers as well as the degradation of the polymers, which are affected by the mixing process, can strongly influence the processing behavior during the extrusion process. For several decades extrusion has become a very important rubber processing technology and thus many detailed research works studied screw and die design. The fundamental investigations assumed rheological and compound properties to be constant. However, the direct influence of the mixing process on the compound properties was not considered in these studies.

As described earlier, the mixing process has a direct influence on both the processing behavior and the product properties. Rheological behavior, filler dispersion, and curing kinetics depend substantially on the mixing process parameters, such as number of revolutions, ram pressure, process duration, and addition sequence of the compound components.

In order to determine the influence of the mixing process on the extrusion behavior mixing processes of different designs were studied. The direct influence of the various process conditions on the processing behavior and the reported part quality were analyzed for a tire compound manufactured by:

- 2-stage and 3-stage mixing processes,
- standard and up-side-down processes.

The difference between standard and an up-side-down mixing sequence is represented in Fig. 5.41.

The up-side-down process resulted in an unfavorable and insufficient dispersion level, which can lead to a higher than standard compound viscosity (see Fig. 5.42).

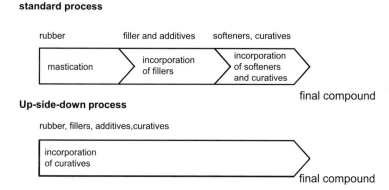

Figure 5.41 Standard and up-side-down sequence

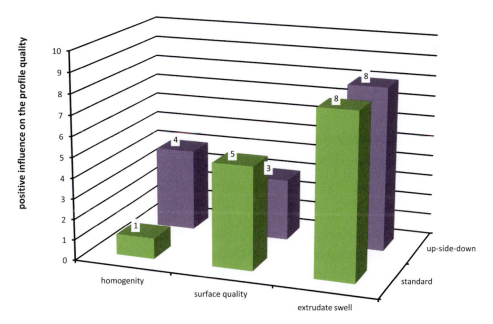

Figure 5.42 Influence of mixing process (standard and up-side-down) on profile quality (higher number indicates better qualities)

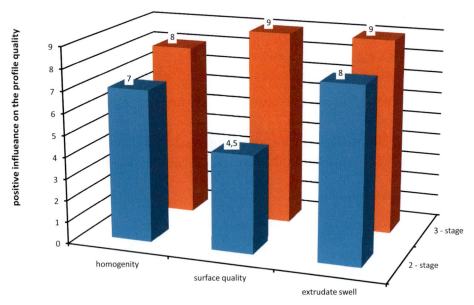

Figure 5.43 Influence of the mixing process (2- and 3-stage) on the profile quality (higher number indicates better quality)

As discussed in Chapter 2, the addition of fillers and oils has a strong influence on filler dispersion, which in turn is directly related to the surface quality of the extruded profiles.

The 2-step process resulted in an increased memory/elastic effect (Fig. 5.43) of the profile which can be attributed to poor incorporation and destruction of the fillers caused by lower mixing energy input.

5.12.4 Influence of the Milling Process on Compound and Part Properties

During the production of processable rubber compounds several rolling mills often follow the internal mixer. A brief description of the rolling mill, its function and purpose within the entire production line will follow here.

A rolling mill consists of two parallel, horizontal rolls rotating in opposite directions, which are driven by single drives or by a gear box. For process engineering reasons it is necessary to cool the rollers independently, especially to modify the contact temperature between rubber and rolls and thus the sticking/sliding conditions. Good cooling can be achieved with peripheral drilled rolls. Here, cooling channels are placed directly under the milling surface. In order to achieve a faster change of friction between the front and rear roller in reverse and during the working process, working with single drives is favorable. By moving the front roll, the gap is adjustable. Two V- shaped chucks are mounted above the rolls, one towards each end of the nip, in order to guide the material into the roller slit and to prevent the flow of rubber mix laterally along the rolls. Beneath the rolls a removable tray catches any small compound parts that may fall through the roller gap or off the roll surfaces. The drive power of roll mills is approx. 2 kW per kilogram of batch weight. Manual blending of the compounds on the roll mill, in order to achieve additional compound homogenization, can be facilitated by a "Stockblender", placed above the rolls. A pair of drawing rolls on top of the front mill drags the complete rough sheet upwards. Two guides before the drag rolls interfold the sheet before the "strand" is fed back to the roll gap. A lateral movement of the guides helps to distribute the compound to the left/right end of the mill.

The Purpose of Roll Mills within the Mixing Process

The compound is discharged from the internal mixer as a shapeless hot mass and drops onto the rolling mill. The pieces of compound are supplied to the mill by either a conveyor system or by free fall directly from the mixer. The pieces of compound are milled to form a band of rubber, which can then be cooled down and homogenized. If the mill follows (as in this procedure) immediately after the internal mixer, the mixing times of the compound on the rolling mill are relatively short. Because the compound covers the entire working length of the roll and thus the cooled surface area of the roll is very large compared to the compound volume, more rapid cooling can be accomplished.

Although milling can have a crucial influence on both compound and part properties, its impact is overlooked and often even neglected during process evaluations. Using sensor system to facilitate extensive monitoring is essential. An example of such a control unit is shown in Fig. 5.44.

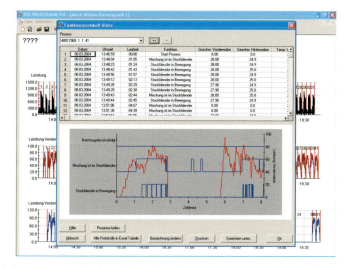

Figure 5.44 Process control of a rolling mill – *Scarabaeus system*

Monitoring of the process parameters, such as electrical power consumption (red), and step control (blue), and the number of stockblender movements is of crucial importance for direct process evaluation by the machine operator and/or the procedure technician.

The rolling process, together with the mixing process in the internal mixer, can be explained in terms of a superposition of the drag flow rate and the pressure flow rate.

Combined they have a clear compounding effect on the rubber compound, which is proportional to the tensile stresses and the shear stresses in the roller gap.

The shear stress is dependant of the following parameters:

$$\tau = f(T, p, \dot{\gamma}, \eta)$$

The "roller gap pressure p" is an extremely important factor and depends on:
- compound viscosity,
- compound elasticity,
- roller gap,
- peripheral speed, and
- friction.

A closed theoretical description of the milling process is currently not possible.

Figure 5.45 shows a summary of the correlations between the important parameters of the milling process:
- number of revolutions
- milling time
- gap

on the one hand and processing as well as compound and part properties on the other hand.

The influences become clear, when other roller gap adjustments (as often employed in the industry) are used. With a simple calculation it becomes clear, that a reduction in roller gap requires an increase of the speed to 30 m/min; with a speed of 20 m/min a reduction in roller gap from 3 mm to 2 mm is required, in order to achieve the same number of gap passes of the compound. This again leads to increasing temperature due to the higher shear stresses and can ultimately result in a reduction of incubation and curing time. Premature scorching of the compound could cause processing difficulties during further processing.

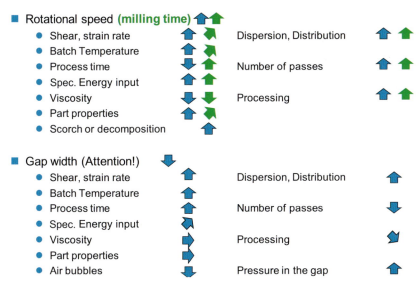

Figure 5.45 Milling process - summary (influence of changes in parameters - increase in number of revolutions and/or reduction of the gap; green - increase of milling time)

$$A_s = L \times s$$

where

A_s – gap area

L – gap length

S – gap clearance

$$\dot{V} = A_s \times v$$

where

v – rotational speed (mill)

$\dot{V}$ – volume flow

$$\text{number of revolutions} = \frac{\dot{V}}{V_{comopund}} \times t_{roll}$$

where

$V_{compound}$ – volume of the batch

t_{roll} – mixing time on the rolling mill

The following example from a production line illustrates this. The monitoring unit for an SBR compound indicated significant differences in the properties of the final parts, in particular uncured areas in the rubber part. A continuous analysis of the

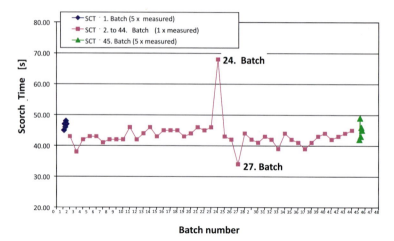

Figure 5.46 Curing behavior of 45 loads of an SBR compound (scorch time)

production process showed that the curing behavior of two batches showed clear differences (see Fig. 5.46).

In order to determine the causes for the property deviations, both the mixing process in an internal as well as the rolling mill process was analyzed. A comparison of the process data for the internal mixer showed a good level of consistency and did not exhibit any deviations. Further analysis concentrated on the rolling mill and the diagram of the rolling mill data (Fig. 5.47) confirmed that the process deviations were due to milling process variations for batches 24 to 27.

Figure 5.47 shows that an interruption of the rolling mill process resulted in insufficient energy input for batch 24, which caused incomplete incorporation of

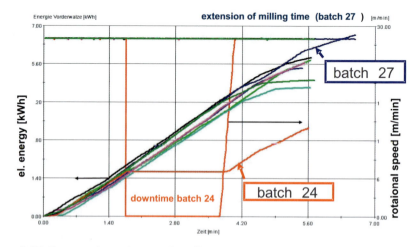

Figure 5.47 Power and speed curve of a mill

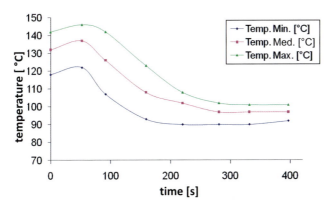

Figure 5.48 Compound temperature gradient on the rolling mill

curatives and a lower temperature and longer scorch time. On the other hand, batch number 27 was mixed longer than planned. This resulted in an increase in energy input, which led to reduction in curing time. The factors energy input and prolongation of mixing or reduction of the curing time were identified as key causes of the processing problems. Poor or missing process control of the rolling process would have made it more difficult or even impossible to identify the causes for the production problems. The mentioned cooling effect of the milling process played an important role here. Figure 5.48 shows the development of the compound temperature on the rolling mill. After an initial reduction in temperature, the compound temperature later changes (only insignificantly) due to the supplied mechanical energy (compound energy). Therefore, extending the milling process causes not only a lower cooling effect in many cases, but can additionally reduce the scorch time due to the thermal effect and heat history.

The Batch-Off

During the production process from internal mixer to roll mill, the compound sheet is continuously cut by knives attached to the downside of the front roller and then taken up to the cooling equipment, to the "batch off". The process is described by the principle sketch in Fig. 5.49. The rubber sheet is dipped into an anti-stick solution, dried, cooled, and then stacked. The rubber sheet is supplied to the batch-off system by a conveyor belt.

The sheet passes through an anti-stick water solution in order to reduce surface tackiness, travels on transverse staffs into endless loops and finally through a channel with large laterally attached fans for simultaneous drying and cooling of the rubber. Figure 5.50 shows that the batch-off unit has an influence on the temperature history of the rubber and thus on the curing behavior of the compound. The influence of the cooling agent and the length of the batch-off unit are shown in Fig. 5.51.

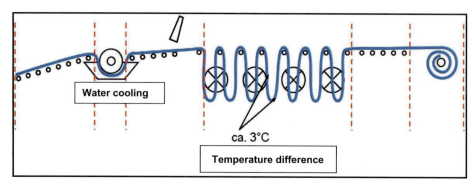

Figure 5.49 Batch-off (run from left to right) – with temperature testing points

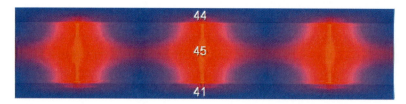

Figure 5.50 Temperature distribution on a rubber sheet (fans below; direction – to the right)

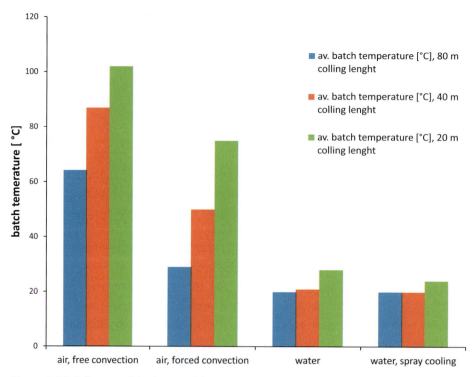

Figure 5.51 Influence of the length of the batch off and the cooling agent on the final temperature of the compound (initial temperature: 120 °C)

The cooled down rubber sheet is is taken off and deposited endlessly into loops (wig wag system) at the end of the machine, or cut and stacked onto pallets, or cut into strips and then deposited into boxes. The cooling process must be as effective as possible. The cooling procedure of low heat conducting compounds in particular can take a long time, therefore the design of the batch-off unit is very important for to the total mixing process. Only when batch-off unit is designed properly, it is possible to realize an appropriate residence time for the short working times of the internal mixer.

■ 5.13 Summary

As mentioned earlier, it is very difficult to maintain accurate compound properties and the desired level of product tolerances using batch mixing processes. Therefore compound characterization methods and constant process control are of crucial importance in actual production systems. The main target is to select the accurate compound characterization methods as well as process parameters that can be supervised and matched with the production process and the product properties. Progress in terms of compound characterization is expected by extensive and broad investigations of the correlations between the production processes (mixing, extrusion and injection molding) and the respective compound and part properties. In order to solve this complex problem, it is important to analyze and then determine the properties of the products (and/or compound) after each processing step. The main focus in this chapter was to give an overview of the correlations of the complex processing chain within the processing plant and also to suggest further analysis and examination of the whole process. A suitable compound testing method, which should match the respective parts as well as the process analysis methods can lead to a reduction of necessary off-line tests and of waste parts in production. In addition, a detailed analysis of the relevant process parameters is a strong measure to optimize production processes; it also allows faster identification of main causes for any production problems. From the manufacturer's point of view, this can be used as basis for economical and property-related process optimization and/or process design.

References for Section 5.8 to 5.13

[1] Röthemeyer, F.: Entwicklung und Erprobung eines Messgerätes zur Erfassung des Verarbeitungsverhaltens von Kautschukmischungen, *Kautschuk + Gummi, Kunststoffe,* **33** (1980) 12, p. 1011–1015

[2] Menges, G., Grajewski, F., Limper, A., Weyer, G.: Bessere Qualitätskontrolle durch Nachrüstung von Standardprüfgeräten, *Gummi Fasern Kunststoffe,* **39** (1986) 3, p. 126–129

[3] Busch, E., Groth, V., Pogatzki, V. W., Potthoff, A., Wetzel, H. E.: Prüfgerät für den Vulkanisationsgrad, *Kautschuk + Gummi, Kunststoffe,* **42** (1989) 9, p. 782–787

[4] Levin, N. M.: Demands on Testing and Quality Technique in the 1990, *Polymer Testing,* **9** (1990), p. 315–327

[5] Burhin, H. G.: Improved Techniques for Charakterization of Polymer and Compounds before, during and after Cure, *Kautschuk + Gummi, Kunststoffe,* **47** (1994) 4, p. 262–267

[6] Michaeli, W.: Das Mischen von Kautschuk und Zuschlagstoffen, in: Der Mischbetrieb in der Gummiindustrie, VDI-Verlag, Düsseldorf, 1984, p. 5–23

[7] Dannenberg, E. H.: *Ind. Eng. Chem.* **44**, (1952), p. 813

[8] Boonstra, B., Medalia, A.: Effect of Carbon Black on the Mechanical Properties of Vulcanizates, *Rubber Chem. Technol.* **36** (1963), p. 115

[9] Payne, A.: Effect of dispersion on the dynamic Properties of Filler-Loaded Rubbers, *Journal of App. Polym. Sci.* Vol **9** (1965), p. 2273

[10] Hess, W. M., Wiedenhaefer, J.: Product Performance and Carbon Black Dispersion, *Rubber World,* **9**, 15 (1982)

[11] Hess, W. M., Swor, B. A., Micek, EG.: Influence of Carbon Black, Mixing and Compounding Variables on Dispersion, *Rubber Chem. Techn.* **57**, (1984), p. 959

[12] N. N.: A Review of European Rubber Research in Practice, International Conference 2002 in Paderborn, Germany, Mini Derucom and PRODESC, KTP, Univ. Paderborn (Eds.), Institut für Kunststofftechnik (KTP) – 2002, 1st edition

[13] N. N.: Prüfungen von Werkstoffen für die Elektrotechnik, DIN 53 482, (1983)

[14] Göttfert, A., Reher, E. O., Thomas, U.: Rheometrie von Elastomeren, brochure Göttfert, 1998

[15] Göttfert, O.: Das Rheovulkameter, ein Prüfgerät zur schnellen Aussage über das Verarbeitungsverhalten von Kautschukmischungen, *Kautschuk + Gummi, Kunststoffe,* **35** (1982) 10, p. 849–857

[16] Nijman, G.: Vredestein Banden B. V., Personal communication, Enschede/The Netherlands 2000

[17] Bußmann, M.: Fa. Freudenberg Dichtungs- und Schwingungstechnik KG, Personal communication, Weinheim 2000

[18] Moos, K.-H.: Möglichkeiten und Grenzen der praxisnahen Beurteilung der Verarbeitbarkeit von Elastomermischungen mit vereinfachten rheologischen Prüfungen. In: Mischungsherstellung – der Kernbereich der Elastomerverarbeitung, Düsseldorf: VDI-Verlag 1997

[19] Vennemann, N., Lüpfert, S.: Mooney-Relaxationsprüfung mit FFT-Analyse eine neue Methode zur schnellen rheologischen Charakterisierung von Kautschukmischungen, *Kautschuk-Gummi-Kunststoffe*, **44** (1991) 3

[20] N. N.: Vulkametrie, DIN 53529, (1983)

[21] Kramer, H.: Rheologische Messungen und Verarbeitungsprüfungen an Rohkautschuken und Mischungen, Bayer AG, KA F+E brochure, August 1996

[22] N. N.: User manual "Rubber Process Analyzer" (RPA 2000), Fa. Alpha Technologies, 1996

[23] Ryzko, P.: Prozessüberwachung und -optimierung in der Elastomerverarbeitung auf der Basis statistischer Versuchsmethodik und mathematischer Prozessmodelle, Dissertation RWTH Aachen, 2001

[24] Limper, A.: Verfahrenstechnik der Kautschukverarbeitung, Von der Mischungsherstellung bis zum Endprodukt, am Beispiel der Extrusion, habilitation treatise RWTH Aachen, 1991

[25] Haberstroh, E.: Kautschukverarbeitung, Lecture notes RWTH Aachen, 1996

[26] Schmid, H.-M.: Qualitäts- und Produktivitätsverbesserungen im Innenmischer, Beitrag in: Der Mischbetrieb in der Gummi-Industrie, VDI Gesellschaft Kunststofftechnik, Düsseldorf, 1984

[27] Ryzko, P.: Einfluss der Mischungsherstellung auf die Bauteileigenschaften, paper VDI/DIK Workshop with practical Tests, Freudenberg 2003 to 2009

[28] Posner J.: Processability tester: A new method for measuring the processability of a rubber compound, *Kautschuk, Gummi, Kunststoffe* **56** (2003) 4, p. 149–158

6 Dispersion and Distribution of Fillers

R. H. Schuster

The diversified and sophisticated property set of modern elastomers can be achieved by blending polymers having specific physical and chemical properties with particulate reinforcing fillers that are necessary to crosslink the entire system to form homogeneous networks. The incorporation of active fillers into a rubber or rubber blend generates unique improvements in physical properties of elastomers, termed "reinforcement" [1 – 3]. It is generally recognized that the main parameters of fillers governing their reinforcing ability in rubber are:

- the size and distribution of filler particles,
- the shape and distribution of filler aggregates,
- the surface activity, which refers to the ability to interact with polymers.

The aim of mechanically mixing the polymer with solid fillers, processing oils, antioxidants, curing agents, plasticizers and others is to cost-efficiently produce a homogeneous mix with filler particles that are reduced as much as possible in size and randomly distributed with the compound ingredients [4 – 7]. Such a mix should demonstrate suitable rheological properties for subsequent processing (extrusion, shaping, injection molding) and the morphological structure that will yield the required physical properties. Considerable time and effort have been directed in developing suitable mixing procedures and the use of proper mixing geometries to disperse the filler in an optimum way [6].

The mixing of rubbers with solid fillers is an energy-intensive process during which maximum dispersion and a good distribution of the raw materials and additives can be accomplished. The process implies breaking down of recipe components – originally supplied in large lumps, bales or granule form – into microscopic volume elements within intermediate or finished compounds. Because solid fillers, i.e., carbon black (CB) [8] or silica [9], are usually delivered in pelletized form, the mixing process should guarantee the break-down of the pellets and then achieve an efficient dispersion of the pellet fragments into small nano-scale particles. Additionally, the particles should be distributed homogeneously in as short a mixing time as possible. The amount of energy required for mixing and dispersing highly viscous polymers into homogeneous blends is mostly realized via mechanical means by suitable mixing tools such as the internal mixer and the two-roll mill. However, the pellet fragments and the large agglomerates present in final mixes significantly influence the ultimate properties of crosslinked compounds. For some applications the filler dispersion should be as good as possible. For other applications, such as improving resistance to crack propagation, an inferior dispersion may be preferable. The filler dispersion is therefore a very, if not the most important parameter to be controlled during mixing. Consequently, precise but rapid characterization methods describing the morphological state of the mix are highly desirable. Though much work has been done in this field in more than eight decades, this topic has not been fully resolved or optimized yet [10].

Improving of physical properties by incorporating particulate filler in polymers implicates the formation of an interface coupling the rigid filler to the elastic polymer phase. The key parameter in the mechanism of interface formation is the filler-polymer interaction that provides phase bonding and strength of the filled composite. While the type and nature of each of the compounding ingredients contributes to the degree of filler-polymer interaction, the development of an appropriate interface interaction during mixing is strongly influenced by the mixing device, mixing conditions, and the compounding ingredients. In order to achieve effective reinforcement by filler particles, a large number of filler-rubber interfaces has to be created by size reduction of the filler particles during mechanically mixing.

The scope of this chapter is to provide an overview of the mechanisms of filler dispersion, methods for determination of filler dispersion, the governing material parameters (type of the filler and the nature of the polymer), the effects of filler dispersion on viscoelastic, dynamic-mechanical and ultimate properties, and finally the filler distribution in rubber blends. Results from the literature and from research in our own laboratory will emphasize the major material dependent contributions to filler-rubber interactions and some routes to improve filler dispersion and distribution by changing the surface activity of the filler as well as the chemical nature of the rubber by specific functionalization. Taking into account that carbon black (CB) – today and probably in the near future – is the most important reinforcing filler used in rubber technology, the majority of the examples refer to this material.

■ 6.1 Dispersive and Distributive Mixing

The general purpose of any mixing process is to disperse and distribute the recipe ingredients into a macroscopically homogeneous rubber mixture. In addition to highly viscous polymers, a multi-phase system also contains physically different substances, such as solid, non-fusible fillers, low-viscosity, soluble liquids (oils), and less soluble, solid additives. It is only on the condition that the well dispersed compound ingredients are evenly distributed that the elastomer material formed by vulcanization will later feature the full scope of properties desired by the materials engineer. It is convenient to distinguish between two essentially different mixing mechanisms [4, 6, 7]:

- dispersive (or intensive) mixing,
- distributive (or extensive) mixing.

Filler dispersion of has been known to be critical to the final characteristics of rubber compounds. During the last decade the subject has gained much interest in the context of development of high performance elastomers, the more efficient use of raw materials, the reduction of rejects, and others. Consistent observations made in the literature support that good filler dispersion improves tire properties, namely rolling resistance [11], tread-wear, and traction [12]. In this context filler dispersion becomes one of the major contributions to product quality and can therefore be used as a quantitative measure to evaluate the quality of a mix as well as to predict product performance. Because currently full understanding of dispersion and distribution is still lacking, the work reported here represents an attempt to provide useful information, taking into account the influence of the morphology and surface activity of the filler, the chemical nature and the molecular weight of the polymer, the type of mixer, and the processing conditions employed during the mixing process.

6.1.1 Dispersive Mixing

Dispersive mixing involves the breakdown (rupture) of the solid (or liquid) constituents from their original sizes into smaller entities of the mix. During the entire process of dispersion the contact surface between the filler particles and the rubber increases (Fig. 6.1). The dispersive mixing can – with the support of mechanical stress and physical-chemical interactions – go as far as to retain molecular disperse systems in borderline cases. This applies to low molecular weight substances such as oils, resins, antioxidants and, in rare cases, to polymers usually degraded to small domains, if sufficient interaction forces come into play. Similarly, it also applies for insoluble solids, such as CBs or silica, which are built up from colloidal units (aggregates) and delivered in the form of pellets. During mixing, the filler pellets (initial size 1–3 mm) are fractured and broken-up by mechanical stress into pellet fragments and large agglomerates (with diameters ranging between 3–5 µm and 500 µm). Subsequently, these "large objects" are more or less degraded in the shear field provided by the rotors of an internal mixer to smaller entities forming the filler aggregates as the smallest solid mono-units (20–1000 nm). As a result of dispersive mixing, the interface becomes larger and phase bonding is improved.

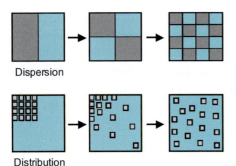

Figure 6.1 Schematic representation of dispersion and distribution

The benefits of dispersion and phase bonding have been experimentally observed in almost all polymer composites; they include higher elastic modulus, improved tensile strength, reduced abrasion, increased cut growth resistance, among others [13]. The specific energy spent to disperse a unit mass of filler in any mixing device is reflected by the process costs, which can be high and, in some cases, comparable to the purchase price of the filler. Therefore, it becomes more and more important to establish the most economic and effective mixing conditions for dispersing the filler to achieve the properties required for a specified application [11]. Detailed knowledge about the material property changes and the state of dispersion in each phase of the mixing process is therefore required.

The inability to achieve a good level of dispersion and consequently the necessary phase bonding in the composite material impairs the ability to realize the full performance of the filler. Due to the fact that filler loading is typically high in many composites and especially in most of the rubber products (≥ 60 phr) and because the mixing time is kept as short as possible, the filler dispersion often does not reach an optimum value [14]. Because of the mechanical dispersion occuring in the mixing machines, part of the filler is always retained in the rubber matrix in the form of pellet fragments and large agglomerates, i.e., an ideal level of dispersion is rarely reached by conventional mixing techniques.

Sub-optimal dispersion creates undesirable characteristics for the end use of the materials. The portion of non-dispersed filler can be regarded as a trigger for material failure under static or dynamic stress [15]. The importance of mixing is underlined by the fact that there are practically no possibilities to rectify the inadequate state of dispersion during the subsequent processing steps (i.e., extrusion, calendaring). It should be noted that the process of dispersion, distribution or reorganization of the filler particles takes place before the curing process is accomplished.

Dispersive mixing is performed mainly in internal mixers equipped with intermeshing rotors or on the two-roll mill. This way, a certain degree of distribution and homogenization of the compound is always attained. With solid fillers, wetting activity and dispersion to finer agglomerates is normally accomplished in internal mixers. Under normal production output conditions, no significant improvement to filler dispersion is achieved on extruders. With regard to the type of flow necessary to sustain dispersive mixing it was experimentally established that extensional (elongational) flow and repeated folding processes will increase the efficiency of dispersive mixing to a higher degree than shear flow [16]. Accordingly, the degree of filler dispersion no longer depends upon stress only, but more so on the total strain imparted to the matrix. Assuming the total strain or the extensional flow is a key variable in dispersive mixing, which finally provides higher degrees of filler dispersion, the question is how to improve the design of the mixing equipment so that increasing efficiency in dispersive mixing can be achieved.

6.1.2 Distributive Mixing

Through distributive mixing a given degree of composition uniformity throughout the mixture is achieved. In an "ideal" distributive mixing process, no particle breakdown takes place and the interface per unit volume remains constant [7]. The randomization of the filler particles with a constant size is the important part of this type of mixing and leads to a homogeneous distribution of the particles. The filler concentration fluctuations in small volume elements of the rubber matrix are decreased during the process, while the average particle size remains unchanged (Fig. 6.1). At length scales much larger than the particle size, the entire system appears as homogeneous. Therefore, homogeneity of a mix is not primarily a matter of the filler particles' dimension. However, it is more beneficial that finely dispersed particles – not coarse ones – are evenly distributed in the mix. The resulting degree of homogenization has a positive effect on further processing as well as on the ultimate properties and lifetime under dynamic loading of the finished products. Complementary to the distributive mixing due to folding and shearing in the internal mixer, distributive mixing can be, and is, continued during material transportation on downstream equipment such as extruders [7].

Polymer blends are frequently used in rubber technology to combine the properties of raw polymers (i.e., oil resistance with aging resistance). Usually the unlike polymers are not miscible and form heterogeneous two-phase blends. For such blends the distribution of the filler in the discrete polymer phases is of great importance. Taking into account that the polymers may exhibit different interactions with the filler, a selective partition or concentration of the filler in the discrete polymer phases can occur. This can lead to a selective reinforcement of one polymer phase with beneficial consequences on the level of certain physical properties. By employing master batch mixing, the distribution of the filler and the filler transfer can be controlled.

However, the requirements for *effective* mixing must be considered equally important as those for dispersive and distributive mixing throughout the mixing system. Because different flow fields may favor the occurrence of one or the other mechanism, the mixing equipment plays a major role in giving one of these mechanisms a more dominant role.

The importance of efficient dispersive and distributive mixing becomes impressively clear in the following summary of detrimental aspects of poor mixing [17]:

- impaired processing and manufacturing uniformity, waste of raw materials and additional energy consumption
- reduced product life, poor performance during service and
- impaired product appearance.

Dispersive and distribute mixing require the optimal use of processing parameters, such as mixing time, rotor speed, fill factor, ram pressure, and others. These parameters have to be adjusted for the compounds under consideration in order to exploit favorable interactions between the compounding ingredients at the molecular level that can lead to the formation of polymer-filler interfaces. Besides these parameters, the nature of the polymer used, its molecular weight, and the surface activity of the filler (energy site distribution, surface roughness), the particle size and structure, all can play a major role.

6.1.3 Quality or "Goodness" of Mixes

If the "homogeneity of composition", which is usually the aim of mixing, is taken to be synonymous to "constant composition in all parts no matter how small the parts", this aim would never be attained at the molecular level, because of the very small size of the aggregates. The degree of homogeneity depends on the scrutiny to which it is subjected. For practical purposes, the scale of scrutiny can be defined as "the minimum size of the regions of segregation that would cause the mixture to be imperfect for the intended purpose" [18]. The avoidance of undesired features was the driving force for developing both better dispersing fillers and more efficient mixing technologies. In addition, considerable effort has been directed to developing suitable mixing procedures, using proper mixing geometries, in order to disperse the filler in an optimum way. Compromises have been made in order to achieve a good balance between the dynamic properties of the compound and the costs of processing.

The quality of a rubber mix is related to the degree of dispersion and uniformity of the particle distribution. By dividing the mixture into a number of portions that are large relative to the size of the filler particles (i.e., each portion contains a substantial number of particles), analytical results of the mixture reveal a mean equivalent to the overall composition. Practically, a mixture with a statistically random distribution of sufficiently small parts is considered a perfect mixture. The "goodness of mixing" of an actual mix is usually rated by how close it comes to the statistically "perfect mix" in terms of uniformity. In approaching this problem, several criteria have been proposed to describe the quality of a mix. One approach considers the ratio between the standard deviation of a "perfect" mix and that of the mix under consideration [19]. The mixing quality index $M_{(qual)}$ can be defined as follows:

$$M_{(qual)} = s/\sigma_r \tag{6.1}$$

where σ_r is the standard deviation of the perfect mixture and s is the standard deviation of the actual mix.

The role of homogeneous distribution of the filler is clearly described in Eq. 6.1. It shows that the smaller the standard deviation is, the smaller the quality index becomes. Alternatively, the homogeneity of a rubber mix can be expressed by using the variance instead of the standard deviation [20]. Marker substances as indicators for the quality of the mix have to be defined and methods for precise and rapid determination of the markers have to be developed.

$$M'_{(qual)} = s^2/\sigma_r^2 \tag{6.2}$$

For the purpose of quantifying the "goodness" of mixing it is useful to consider a complex frame of properties: the dispersion index of the filler, the size distribution of the filler particles, the local distribution of the filler, and the partition of the filler in the discrete phases of a polymer blend. Currently, these values are determined by time-consuming off-line procedures. Attempts have been made to establish on-line characterization of the distribution of elements such as zinc or sulphur in rubber mixes by using laser induced spectroscopy. [21].

■ 6.2 Mechanism of Filler Dispersion

On the one hand, the spatial distribution of the components, the degree of pellet break-down when processed in the mixing device, and the detailed topological structure of the mix depend on the type of mixer and mixing operation conditions and on the other hand on the nature of the compound constituents. The former set of influences provides the necessary mechanical stress and energy to disperse and distribute the filler particles. The latter set of properties determines the material response to the favorable conditions provided in the mixing process. These properties govern the polymer-filler interaction and come into play when the average size of filler agglomerates becomes small. For a given formulation, in which the interaction between the filler and the polymer ("wettability") is essential, the further erosion of the pellet fragments and de-agglomeration are critical processes for good dispersion.

6.2.1 Theoretical Approach

The role of mixing is to apply a mechanical stress and/or strain, which is higher than the cohesive forces that hold aggregates together inside an agglomerate. In this regard, a model for agglomerate break-up [22, 23] describes the role of hydrodynamic stress in a simple shear flow applied to an agglomerate which consists of a variable number of aggregates (the solid mono-units and discrete rigid colloidal entities). Commonly, the problem is reduced to the breakdown of a two-particle agglomerate in the shear field [23], which generates a hydrodynamic drag force acting on the agglomerate by separating it at its weakest link. Assuming an average connection number v_F with which an aggregate is bound to the agglomerate and the mean interaction force of a connection H, the cohesive force, F_c, for an aggregate resisting separation from the agglomerate is:

$$F_c = H \cdot v_F \tag{6.3}$$

where H is the mean interaction force and v_F the number of connections.

The hydrodynamic force F_H acting to erode the agglomerate is a function of the medium viscosity η and the shear rate τ_{xz}, the size of the agglomerate R_a and the portion of the agglomerate surface exposed to the shear field:

$$F_H = c \cdot \pi \cdot R_a^2 \cdot \tau_{xz} \tag{6.4}$$

where c is a function of the size of the aggregate relative to the agglomerate.

The rupture of the aggregate from the agglomerate cluster takes place, when the hydrodynamic force F_H overcomes the cohesive force F_c and it holds:

$$(\eta \tau_{xz})_{crit} \geq \frac{H v}{c \pi R_a^2} \tag{6.5}$$

For the same size of agglomerate the critical rupture stress increases with the inter-aggregate forces and the number of binding contacts per aggregate. For the same bonding energy, the rupture stress decreases with the size of the agglomerates R [14]. The critical size of agglomerates R_{crit} is the size of non-dispersed agglomerates representing the "macro-dispersion" (see below). If the agglomerates' radius is large compared to the cohesive forces (filler-filler interaction), the drag forces overwhelm the cohesive forces and the dispersion proceeds efficiently.

Therefore, the process tends to reach a limit when agglomerates become smaller. Asymptotically the process tends to reach a limit, when the cohesive forces reach the same order of magnitude as the drag forces; when the agglomerates' size is below a critical value. With all advantages of the model describing the influence of rheological properties on filler dispersion, the question of understanding how shear forces can be transmitted from the polymer melt to the outer part of the agglomerate remains unanswered.

6.2.2 Phases of Mixing Process

When pelletized fillers are mixed into any polymer melt, first they have to be incorporated into the polymer phase [4, 24]. In the initial phase of the process the filler pellets are broken-up into pellet fragments under high shear forces. Therefore, it is important to find the optimum conditions to efficiently break the pellets in as small fragments as possible by properly selecting the mixer, rotor design, and the process parameters (see Chapter 2). Pellet rupture is facilitated by process parameters such as rotor speed, cooling, fill factor, and ram pressure. It follows from Eq. 6.5 that large pellets or/and agglomerates are easier subjected to fragmentation and rupture than small ones. During the incorporation phase the "wetting" of the solid surface and encapsulation of the pellet fragments by the polymer provide the starting platform for further size reduction. Due to shear forces and pressure in the mixer, the polymer is squeezed into the void spaces of the filler agglomerates, replacing air [25]. At this stage, a "venting" of the mixer becomes necessary. By replacing the air from the voids, the volume of the mix is significantly reduced and the specific gravity is correspondingly increased. The reduction in volume is considerable, leading to a reduction in torque and power consumption (see Section 6.5.3). As a result of the filler incorporation, the mix does no longer contain "free" filler and is not emitting any dust. Because of the energy required to overcome the cohesive forces within filler agglomerates, the "dispersion" phase is the main power consuming phase during the mixing process. The size reduction takes place under high shear and extensional flow and consists in a gradual erosion of the outer layers of the agglomerates. By decreasing the size of the agglomerates the critical agglomerate radius R_{crit} (see Eq. 6.5) is approached. From this point on, the flow field provided by the rotors is transmitted to the outer layers of the agglomerates by the polymer-filler interface.

The dispersion phase is qualitatively described by the "onion skinning" model presented by Shiga and Furuta [26]. The small agglomerates and aggregates sheared off from the pellet fragments are distributed in flow direction around the pellet fragments. The model explains that the number of small agglomerates and aggregates increases exponentially during the dispersion phase. Characteristic for the dispersion phase is an increase in torque, power consumption, and temperature until a maximum value is reached. Due to the increasing particle number, the inter-particle distance approaches a critical value at which filler network formation starts. This can be easily observed in case of electrically conductive fillers (i.e., CB). When the filler network is formed and the electrical conductivity threshold ("percolation") is reached, a significant decrease in the resistivity of the mix is observed. By prolonging the mixing process the dispersion phase reaches its limit, because the temperature increase would normally lead to a viscosity drop and shear forces tend to decrease to a minimum value. However, the "onion skinning" or erosion of agglomerates is still active or predominant at this stage of the mixing process. At the end of the dispersion phase a certain particle distribution is reflecting the efficiency of these complex interactions in the mixer [27]. The process of dispersion and distribution described by the "onion skinning" model is schematically shown in Fig. 6.2.

Depending on the specific mixer, shear stress, extensional stress and strain as well as concentration randomization of the filler particles can be accomplished by complex pseudo-random flow patterns between the rotors or between the rotor wings and the wall of the mixing chamber. The necessary shear flow appears essentially in the nip regions, where the rotors rotate in opposite directions (similar to a two-roll mill), between the rotor surfaces and the inside surface of the mixing chamber, and between the surface of the rotors that are aligned at a variety of angles to the directions of rotation.

For internal mixers equipped with tangential rotors (which do not overlap), mixing efficiency is achieved in the region between the rotor flights and the walls of the mixing chamber. Better dispersive mixing can be achieved by increasing the number of flights and changing their profile [28]. The advantages of this type of

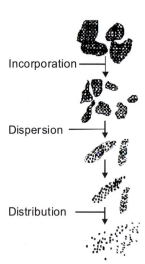

Figure 6.2 Mechanism of carbon black dispersion and distribution [26]

mixer include short incorporation times and the larger batch volume that can be processed.

By contrast, the intermeshing rotor geometry demonstrates a more effective dispersive mixing because of the small inter-distance of the rotor surfaces that guarantees a higher degree of shear and extensional flow during one rotation of the rotor. Another advantage of this rotor geometry is the much higher surface to volume ratio, which allows more efficient cooling than in the case of tangential rotors. The incorporation times are slightly longer in this type of mixers, but the mixing efficiency is significantly increased, especially when quality mixing and micro-dispersion are important. While the intermeshing rotors are frequently used in the technical rubber goods industry, now they become more and more important in tire manufacturing.

With some delay to the dispersion of the pellet fragments the closely spaced agglomerates and aggregates are randomly transported through the mix. The "distribution" phase is supported by repeated shearing and folding of the mix. Due to the randomization, the filler aggregates are separated from each other. Differences in local filler concentration are reduced and concentration fluctuations (see Eq. 6.1 and 6.2) inside the mix are asymptotically reduced. The filler network is affected by the process of departing filler clusters from each other and isolating filler particles by polymer layers. This leads to changes in rheological properties, i.e., a decrease in viscosity and, in case of conductive fillers, a decrease of electrical conductivity. However, even at the end of this stage a certain degree of concentration fluctuations and the presence of pellet fragments are observed in many cases. This is the reason why the standard deviation of a real mix is always higher than the one of an ideal mix, resulting in the mixing quality index $M_{(qual)}$ being less than 1.

6.2.3 Polymer-Filler versus Filler-Filler Interactions

The fact that different rubbers with similar viscosities are dispersing the same type of filler differently cannot be explained by this approach. Therefore, regardless of the rheological conditions, the state of dispersion has to be considered as balanced by two counteracting interactions. On the one hand the polymer-filler interaction is responsible for wetting the filler surface and the formation of an interface. All the contributions to this interaction are improving filler dispersion. Their role is similar to the one of the rheological parameters (viscosity, shear rate, extensional flow) when dispersion at a small length scale (micro-dispersion) is concerned, where shear stress has to be transmitted to small agglomerates. On the other hand, the filler-filler interaction also has to be considered. It increases the cohesive energy of agglomerates and thus works against the former type of interaction. Considering the average size of a filler particle at a given time of the mixing process, the influences can be qualitatively described as follows:

$$\bar{d}_{(t)} = \frac{\varepsilon_{F-F}}{\varepsilon_{F-P}} \times \frac{\gamma}{\eta_m \, \tau_{xz}} \tag{6.6}$$

where, ε_{F-F} and ε_{F-P} are the filler-filler and the polymer-filler interaction energies, respectively, γ is the interfacial tension, τ_{xz} is the shear force, and η_m is the viscosity.

It becomes clear, that the average agglomerate size, $\bar{d}$, which is used as a measure for dispersion, decreases when both the ratio of filler-filler to polymer-filler interaction energy and the interfacial tension decrease and the shear forces and the viscosity in the polymer matrix increase.

Polymer-Filler Affinity
Key to the understanding of wetting and polymer-filler interaction is the knowledge about polymer adsorption on solid surfaces. Adsorption leads to a relatively strong attachment of chains on the surface and to the formation of polymer layers, which irreversibly adhere to it [29]. Therefore, polymer adsorption depends on the affinity between the polymer and the adsorption sites on the filler surface. Generally, the term affinity summarizes favorable interactions between the system constituents. For a polymer-filler system, affinity describes the contributions of polarizability of chain segments, the ability to form weak or strong interaction bonds (van der Waals type, dipole-induced dipole, dipole-dipole, hydrogen bonds etc.) and the contributions of adsorption sites on the filler surface. An important contribution to interaction of polymers chains with others or with energetic sites of fillers is the polarizability of chain segments, i.e., the ability to form dipoles in an electrical field. In special cases, chemical reactions between the polymer chains and functional groups on the filler surface can occur directly or via a coupling agent (i.e., silanization of silica and coupling to unsaturated polymers, especially solution SBR) [9].

The rather complex phenomenon can be understood by considering the polarity of the polymer and the filler. In case of a polarity mismatch (i.e., polarophobic chain segments and polarophilic filler surface), the affinity is reduced and the interface thickness becomes small.

In cases of similar polarity, the affinity is higher and the amount of adsorbed chains is larger. The corresponding interfacial thickness can be relatively large (several nm). A quantitative measure for the polymer interaction potential is provided by the cohesive energy density of the polymer [30]. The value describes the ability of chain segments (or radicals) to establish intermolecular forces between unlike species and is defined by the solubility parameter δ which can be determined experimentally [31] or calculated from tabulated structural increments [32, 33]. The difference of solubility parameters provides a characterization of interfacial tension and is the base for phase compatibility predictions [34]. The δ parameter concept was used to estimate the polymer-filler affinity for non-polar rubbers and CB [35]. Qualitatively it holds, the smaller the δ parameter difference, the higher the affinity between the interacting partners.

Interface

A comprehensive treatment of polymer-filler interactions has to consider the average interactions over the polymer-filler contact surface in the unit volume of the mix. Assuming "ideal" filler dispersion, where the contact surface for fillers exhibit little porosity, the total area of the interface depends on both the loading and the surface specific area of the filler. In a unit volume of compound the interface, Ψ, is given by [36]:

$$\Psi = S \cdot \rho \cdot \phi \qquad (6.7)$$

Where S is the surface specific area, ρ is the density of the filler, and ϕ is the filler volume fraction.

Due to the increasing interest in nanocomposites, the investigation of the properties of polymeric interfaces became a focus in modern material science. In order to explain the rheological behavior and the dynamic properties of CB-filled rubbers, a rubber shell of 2 – 5 nm on the filler surface was postulated [37]. The formation of a more or less immobilized layer surrounding the filler particles results in restrictions of the polymer dynamics in the vicinity of the filler. The model is supported by investigations of nuclear spin relaxation time, which reflect the polymer chain mobility [38, 39]. Three different states of mobility were detected for CB filled BR and EPDM (unbounded rubber, bonded rubber in a shell and tightly bonded rubber). Results obtained by other physical-chemical methods strengthen the model. IR-spectroscopic investigations of bound rubber indicate characteristic shifts of vibration bands or the formation of new interaction bands [40]. Reinforcing fillers can reduce the degree of swelling of the rubber phase by an amount that is related to the filler loading and the surface activity of the filler [29, 41] and ultrasonic spectroscopy [42, 43] supports the existence of a rubber shell on the filler surface. It is safe to conclude that polymer chains within the interface are more densely packed and the chain mobility is strongly reduced. Due to the higher segment density, the physical properties of the interface differ from the polymer bulk properties. Although there is sufficient evidence for the existence of the interface, considerable disagreement exists with regard to the thickness or the volume fraction and especially the question, whether the polymer in the interface has glassy characteristics [44, 45]. From a theoretical estimation of the inter-particle distance as a function of the filler particle size and the volume fraction it follows that very small inter-particle distances of few nm can be achieved for nano-particles (< 30 nm) even at volume fractions of less than 0.2.

Assuming an interface thickness of 5 nm, the volume fraction of the interface in the composite becomes comparable to that of the not adsorbed polymer. Thus, the contribution of the interface increases in particular with the affinity between polymer and in general with the filler and with filler dispersion. The mechanical behavior (modulus, stiffness, hysteresis, tensile properties, etc.) of the composite will be determined by the large interface at small filler volume fractions.

Despite the fact that the existence of the interface and its impact on macroscopic properties of nanocomposites was proven by various methods, it is obvious that Eq. 6.7 describes a theoretical contact surface that has not been entirely established in practice. The difficulties with the prediction (or estimation) of the real contact surface arise from the relatively poor knowledge about the cohesive forces within filler agglomerates, especially of inter-aggregate connection forces F_c and the connection number v_F (Eq. 6.5). Useful information in this respect can be obtained by taking into account the structure of filler aggregates and the surface activity of fillers.

Particle Size

Generally, filler aggregates are composed of "primary" particles fused together into aggregates [46]. These are considered the discrete solid units of the filler that exist in rubber vulcanizates. The size of the "primary" particles and aggregates can be determined from TEM images at large magnifications [47, 48]. The specific surface area (EMSA) of CB can be derived directly from the surface mean diameter, using the expression:

$$S\,(m^2/g) = 6000/\rho \cdot d_{sm} \tag{6.8}$$

where ρ is the density of CB in g/cm^3 and d_{sm} is the mean diameter in nanometer.

The values are useful means of comparing different CBs because they are based on particle size distribution [49]. The surface specific area is determined by the number of gas molecules in a monolayer adsorbed on the filler surface [50, 51]. There is an inverse proportionality between the aggregate size and the surface specific area. For furnace CBs and thermal CB, which posses no low surface porosity, higher surface area blacks have smaller aggregate size. Taking into account that fillers provided by suppliers cover a large range of aggregate sizes or surface specific areas (CB from 9 to 140 m^2/g; silica from less than 70 to more than 190 m^2/g), significant differences in the achievable contact surface at constant loading can be expected.

Particle Shape

CB aggregates vary in shape from spheroidal particles that occur in thermal blacks to branched types common to furnace blacks. TEM investigations have revealed the irregular and branched (fractal) shape of the aggregates of furnace CBs [52]. The branched nature of aggregates creates voids within agglomerates that are formed due to filler-filler interaction. These voids are much larger than those occuring by close packing of spheres. The void volume is characteristic for a particular CB grade [25]. By increasing aggregate branches, the void volume increases and the solid fraction within the agglomerate decreases. Consequently, the average number of inter-aggregate connections v_F within the agglomerates decreases. For CBs with the same surface specific area the number of connections between aggregates decreases as a function of the void volume. Assuming constant force H for each inter-aggregate connection within the agglomerate, the cohesive force that keeps agglomerates intact should be smaller for CBs with branched aggregates (because of the smaller number of connections) and larger for CBs with more compact aggregates (due to more connections). The former should break more easily under the same hydrodynamic stress than the latter. Therefore, an easier dispersion and a higher dispersion rate can be predicted for fillers with large void volumes.

The void volume is commonly related to the filler "structure". The techniques for measuring the "structure" are based on absorption of liquids into the internal voids of agglomerates. The DBP absorption is one of the most commonly used techniques to measure the level of absorptivity of CBs [53]. The "structure" of CBs can be controlled during manufacturing in a wide range, e.g., by addition of KCl [8]. As a result, the DBP-number of CBs ranges from low structure blacks (LS-CB) with ca. 30 ml/100 g to high structure blacks (HS-CB) with 130–160 ml/100. It was proven that the DBP number correlates well with the perimeter fractal dimension [54, 55] and the mass fractal dimension of CBs [56, 57].

Surface Activity
While filler particle size and void volume are determined on the basis of standardized gauging methods, the characteristics of surface activity are less clearly defined and not yet subject to determination by standardized processes. Surface activity encompasses every filler surface characteristic that contributes to polymer/filler or to filler/filler interaction [10]. It ranges from surface topology and roughness, which supports anchoring of polymer chains on the filler surface, to adsorption sites that exhibit a certain interaction potential to the polymer segments or support inter-aggregate connections by attractive forces. In addition, specifically interacting functional groups have to be considered in special cases (e.g., silica). For a better understanding of the phenomenon of dispersion it is helpful to focus on the energetic aspects of surface activity.

Early investigations performed by X-ray scattering revealed the semi-crystalline nature of CBs [58–60]. Quasi-planar graphitic micro-crystallites are randomly arranged on the aggregates' surface. It has been shown that the various types of CB differ in the magnitude of their variation of graphitic micro-crystallites. The size and amount of these micro-crystallites increases when the diameter of the primary particles becomes larger. By annealing the CB at high temperatures in an inert gas atmosphere the quasi-spherical "primary" particles of the aggregates demonstrate almost planar graphitic surfaces [61]. Using Raman-spectroscopy with CB showed that domains of "amorphous" carbon occur around graphitic micro-crystallites [61].

Adsorption studies provide an initial means of gaining information on the energetic state of filler surfaces [62]. By using inverse gas chromatography (IGC) the dispersive and polar components of the surface activity (γ_{sd} and γ_p) were investigated [63]. It was seen that the surface energies of higher surface area CBs, both the dispersive and polar components, are generally higher than those of their lower surface area counterparts. This implies that from the view of surface energies, i.e., the mean force H between aggregate connections is higher and therefore, the dispersibility of fillers with high surface area is lower. From static gas-adsorption of molecules with an analogue structure to characteristic polymer segments, the distribution function of surface energy sites can be determined by applying the idea that the experimental adsorption isotherm is the result of adsorption energies of patch wise distributed adsorption sites [64]. Significant progress was made by employing this method for furnace CB and graphitized grades [65, 66] as well as precipitated silica [67]. The experiments confirmed that on the surface of all CBs manufactured in the furnace process there are four discrete energy sites present. The graphitic micro-crystallites establish the weakest interactions (15 kJ/mole) towards hydrocarbons. More efficient are the sites containing amorphous *(polymorphous)* carbon (20–21 kJ/mole). The next two adsorption sites are edges of crystallites (30–31 kJ/mole), and sites between differently oriented crystallites (35–36 kJ/mole).

More important is the result that the fraction of high energetic sites per surface unit increases inverse proportionally to the particle size. In other words, the more abundant topological "defects" caused by the organization of the micro-crystallites and the amount of amorphous carbon on the surface of CBs with small aggregate size lead to a higher inter-aggregate interaction and provides more adsorption sites for polymer chains to form an interface. These results are underlined by the observations that the three high energetic sites disappear when CB is submitted to graphitization (inert gas atmosphere; 2500 °C). On graphitized CB there are practically only the low energetic sites of graphitic crystals present, but they exhibit a significantly reduced interaction potential compared to amorphous carbon or the edges of micro-crystallites. Therefore it can be concluded that for CBs with small aggregate size the connection forces are higher and the resistance to hydrodynamic stress during the dispersion phase is more pronounced.

The surface roughness associated with the topological arrangement of graphitic micro-crystallites was detected by high-resolution TEM [69] as well as AFM [70] and STM [71]. The roughness observed correlates with the surface fractal dimensions determined by static gas adsorption [72]. It was found that the surface fractal dimensions change insignificantly when the particle diameter decreases over the range covered by furnace blacks (d_s = 2.59 – 2.63). This is a severe argument against theoretical attempts to explain reinforcing mechanisms by differences in surface roughness [73].

Precipitated silica shows a quite different surface energy site distribution than CB. The major part of the surface is covered by Si – O – Si bonds with low adsorption energy [67]. The remaining surface is covered by the more polar silanole groups in higher energetic states. By annealing the precipitated silica the silanole groups are reacting into Si – O – Si groups by the process of water elimination. As a result, the surface energy site distribution demonstrates only a single low energy peak. A similar result was obtained by treating the precipitated silica with mono- and bi-functional silanes. Because in both cases the more polar silanole groups are no longer present on the filler surface, the interaggregate interaction is significantly reduced (Chapter 4) [74].

■ 6.3 Dispersion Measurements

One of the challenges to be addressed is the appropriate measurement of dispersion. Dispersion is a length scale phenomenon that describes the aggregate/agglomerate length and the size distribution from the nanometer up to the millimeter scale. Adequate rules to cover the entire phenomenon have not yet been found. Figure 6.3 indicates the relationship between the length scales at which the dispersion has to be assessed and the instrumentation available. The most common methods and their range of application are indicated in Fig. 6.3.

The most common methods seize the "macro-dispersion" on a length scale larger 1 µm. This limit is dictated by the wave length of visible light used in optical microscopy or the mechanical devices to explore surface topology. The amount of undispersed pellet fragments and large agglomerates is quantified by methods described briefly in the next section. One can appreciate that the smaller the scale to be investigated the more sophisticated the equipment needed to evaluate it.

The investigations on a length scale below 1 µm that target small filler agglomerates and aggregates are more difficult but can deliver valuable information for technical application. Even if the borderline to the "macro-dispersion" is not clear cut, the "macro-dispersion" gives relevant information in the sub-micron region. Different methods can provide useful information about filler network inter-aggregate distance distribution, the size and shape of agglomerates, and aggregates formed at different stages in the mixing process. However, the former methods are straightforward and well established, whereas the latter are still under development.

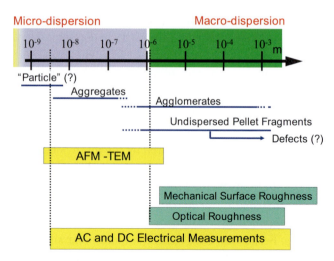

Figure 6.3 Size ranges and methods to measure filler dispersion

6.3.1 Macro-Dispersion

6.3.1.1 Optical Transmission Microscopy

Even before synthetic rubbers were commercialized filler dispersion – as the most important quality characteristic of filled compounds and vulcanized materials – was characterized by the use of surface inspection of stretched, cut, or torn surfaces. The earliest publications relevant to filler-dispersion analysis in rubber were based on light microscope transmission procedures on microtome sections [75–77]. The number of undispersed pellet fragments, their area, and perimeter are calculated and related to the mixing time or the total energy input during mixing (Fig. 6.4).

As can be seen in Fig. 6.4, the mechanical shear forces tear apart the pellet fragments. The elongated dark areas indicate the fragments that are undergoing incor-

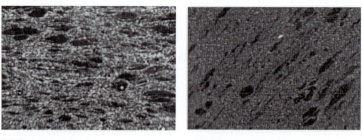

Figure 6.4 Optical microscopy images obtained for 1 and 2 minutes mixing time (magnification 750 times)

poration/dispersion [78]. The almost perfectly rounded areas represent the "hard" fragments. Such pellet fragments will be diminished in number and area during mixing. However, they may be present in the rubber mix even after prolonged mixing.

6.3.1.2 Optical Roughness Measurements

Because of the less involved preparation required, there is widespread application of methods to evaluate the degree of dispersion by comparing the roughness of glossy sections or fracture surfaces with a number of reference specimens. Not very elaborate, the simplest light-optical methods evaluate dispersion from good to poor on the basis of the optic impression of the agglomeration under investigation with the help of a numerator grid [79]. However, there are disadvantageous effects caused by the subjectivity in the assignment into dispersion classes or ratings, the minimal quantification potential and the susceptibility to error of the transmitted-light measurements due to variations in the thickness of the specimens.

The visual inspection of torn vulcanizate surfaces remains popular and is widely used today as a procedure to quantify the extent of filler agglomeration through an optical characterization of the surface roughness. It is convenient to quantify the filler dispersion on the basis of the luminous reflectance of the fracture surfaces [80]. When a surface of rubber compound cut by a razor blade is illuminated under a given angle, the planar and flat regions of the surface containing the polymer and well dispersed filler will reflect the light under the same angle as the incident beam, whereas the convex or concave curved regions of the surface originating from pellet fragments and agglomerates will divert the light under an angle which corresponds to the curvature of the surface [80].

The most versatile mode is to place the light source in the objective of the microscope and to put the light beam vertically on the sample (Fig. 6.5 a) [81]. The size, size distribution, shape, and surface fraction of the undispersed filler agglomerates can be quantified by a computerized image analysis. The method can be applied to any kind of filled-rubber compound. A comparison between a good and poor dispersion and the corresponding quantification is given in Fig. 6.5 b [81].

Assuming that the mean solid content in CB agglomerates is about 0.4 [79] and the swelling for the entire blend is the same as that for the carbon agglomerate, the dispersion index DI can be expressed according to the ASTM standard:

$$DI\ (\%) = 100 - 0{,}4\ V/\phi_r \tag{6.9}$$

where ϕ_r is the volume fraction of filler, and V is the surface area covered by agglomerates.

DI describes the projected area of undispersed filler (pellet fragments and agglomerates) in relation to the area of the matrix that does not contain CB with particles

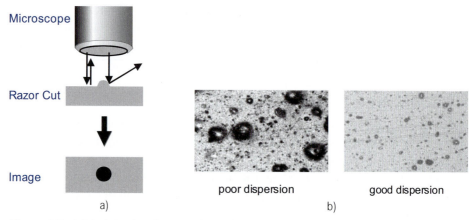

Figure 6.5 a) Principle of surface roughness examination by light reflectance and
b) Optical roughness images for filled rubber mixes (EPDM /CB N 762 /CB 550)

smaller than 3–5 µm. The optical roughness measurement is a user-friendly method that can be performed quickly and delivers reliable and reproducible results about the filler macro-dispersion.

6.3.1.3 Mechanical Scanning Microscopy

In addition to the light-optical method, it is possible to mechanically scan the irregularities on glossy sections caused by the filler agglomerates by using a surface profilometer equipped with a 2 mm diamond tip stylus. The so-called "stylus surface determination" is part of ASTM D 2663 (method C) [82]. The contact pressure generated by the tip on the rubber sample is usually adjusted to minimize the damage to the surface. The results of the measurement are expressed as the difference Ra in the altitude of a point of the surface and the average altitude of the surface (Fig. 6.6). In general we can assume that the smaller the roughness parameter Ra the better the filler dispersion [82].

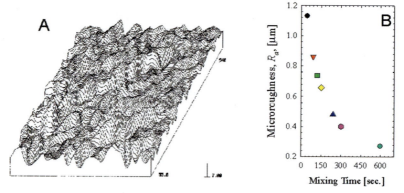

Figure 6.6 Typical mechanical scanning microscopy data a) surface roughness;
b) Ra as a function of mixing time

The advantages of this method include the simplicity of sample preparation and data treatment. It can only be applied to cured compounds and this reduces its usefulness for monitoring dispersion during compound processing in a factory. Compared to the optical roughness measurements, the mechanical scanning microscopy shows lower resolution between samples with comparable high degree of dispersion [83].

6.3.1.4 Reflectometry

In general, CB will absorb the light and the polymeric matrix will diffuse light. Therefore, the general reflectance of visible light on a rubber vulcanizate provides information at the level of the aggregate scale [84]. The decrease in reflectance with increasing mixing time is associated with the balance of the polymer/CB area near the surface of the sample. It is interesting to note that this method can successfully be applied to uncured rubber compounds and thus can be used for monitoring the dispersion level during the production process. It should be stressed, however, that absolute reflectometry data should be used only to compare compounds of the same formulation.

6.3.2 Micro-Dispersion

6.3.2.1 Electrical Measurements

For rubber mixes containing conductive fillers (conductivity of CBs is in the range of $10^{-1} - 10^{2}$ (Ohm · cm)$^{-1}$), tests of electric conductivity and resistance are regarded as sensitive determination methods for filler micro-dispersion. In general it holds: the better the filler particles separation, the higher the resistivity of the compound. When aggregates are separated from each other there is almost no current passing through the sample. By increasing the loading, at some point the percolation threshold is reached and the sample becomes significantly less resistive (Fig. 6.7).

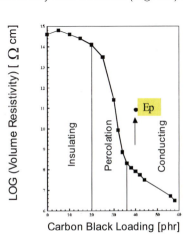

Figure 6.7 Electrical resistivity of a filled rubber compound as a function of CB-loading [54]

Three distinct zones can be observed: insulating zone, percolation zone, and conductive zone [85]. The filler network formation, probed by electrical measurements, is considered an indicator of the degree of micro-dispersion.

The electrical methods are divided into two complementary groups of equivalent importance. One group involves the direct current (DC) while the second considers alternating current (AC) [86]. For the DC experiments usually the current source and a voltmeter is all that is needed, while for the AC method sophisticated instrumentation is required.

6.3.2.1.1 AC Method (Impedance Spectroscopy)

Impedance spectroscopy is a perturbation-response technique whereby the test system (rubber sample) is perturbed (from an initial steady-state condition) by a periodic electrical stimulus and the electrical response is monitored. In general, the relationship between input and response is described by the system transfer function, which contains information about the system under study. The analysis is usually performed in the frequency domain, and the transfer function in this case becomes the electrical impedance [87]:

$$Z^*(\omega) = Z'(\omega) + iZ''(\omega) = V(t)/I(t) \tag{6.10}$$

where, ω is the angular frequency (in rad s^{-1}), $V(t)$ is the periodic voltage perturbation, and $I(t)$ is the corresponding periodic current response.

The resonance frequency can be determined as a frequency at the maximum of $Z''(\omega)$. It has been shown that there is a degree of correlation between the resonance frequency and the dispersion level of CB. AC measurements can be valuable in resolving small differences in filler dispersion and determination of electrical percolation threshold [88].

6.3.2.1.2 DC Method

This technique is applicable to both cured and uncured compounds. Essentially, three types of measurements can be performed by the DC technique, namely volume resistivity, surface resistivity, and current vs. voltage measurements. Simple volume resistivity measurements can provide information on dispersion and can therefore classify the polymers by efficiency of CB dispersion [12].

6.3.2.2 Transmission Electron Microscopy

In face of the early interest on filled rubbers it is not surprising that shortly after the invention of the transmission electron microscopy (TEM) by von Ardenne [89] not only morphological investigations were published but also methods for evaluating the CB dispersion [90]. Preparation techniques were sought for recording the degree of dispersion of the agglomerates using strongly diluted suspensions of rubber compounds [91] as well as embedment in plastics and resins [91], and ultra-thin microtome cryo-cuts (50 – 100 nm). Using pyrolytic or plasma etching [92], it was possible to demarcate polymer phases in filled rubber blends and to evaluate filler distribution. While the determination of filler dispersion becomes difficult at high filler loadings, the method can be applied for small filler loadings. It was confirmed that filler agglomerates can be separated from the rubber mixes by careful extraction followed by morphological TEM investigations. The characteristics of agglomerates, such as size, shape, perimeter fractal dimension, solid content, and the size distribution were evaluated as a function of mixing conditions (Fig. 6.8). The irregular shape of aggregates becomes clear from visual inspection and computer aided data treatment.

Due to the sophisticated sample preparation and the reduced relevance at high filler concentration, TEM is typically employed for research on micro-dispersion.

6.3.2.3 Atomic Force Microscopy

The ability of the AFM technique for mapping a surface by repeated parallel traces delivers complementary morphological data on filler dispersion [93]. Attractive or repulsive forces acting between the surface of a sample and a very small tip placed on a cantilever are recorded when the tip is scanning over the sample surface [94]. For filler dispersion measurements the surface roughness can be recorded in non-contact mode, contact mode, lateral force mode, and in tapping-phase contrast mode above and below the μm scale. The AFM micrographs demonstrate not only the differences between a poor and a good dispersion, but reveal at high resolution characteristics of aggregates and of primary particles and at a low resolution the pellet fragments. The main advantage of AFM is that the sample preparation is easier than for TEM. However, it is still difficult to quantify the degree of dispersion with the AFM technique unless powerful imaging software is employed.

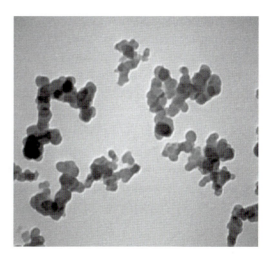

Figure 6.8 TEM images of individual agglomerates extracted from rubber mixes

6.4 Control of Dispersion by Process Parameters

6.4.1 Mixing Procedures

Portrayed in simplified terms, applied mixing methods are subdivided into two groups with regard to the mixes obtained. This subdivision essentially takes into account the incorporation and dispersion of the filler, which is crucial for practical purposes [95].

The first group unites the techniques and mixing methods in which all pure components are mixed in one mixing stage. The filler is added after the rubbers have been premixed for a short time at one or more working steps. These direct methods, are used when a certain distribution of the filler in the polymer phases is not of importance or when filler dispersion is kept below the achievable optimum.

The second group of mixing techniques is targeting the blending of incompatible rubbers and the control of filler distribution. The first step is always the separate production of rubber/filler/oil batches with a different polymer basis, to which crosslinking agents are added at the end of the mixing process. Afterwards, the ready-made batches are blended mechanically in the proportions required for the best performance of the final rubber product. The primary objective of employing these techniques is to control the composition of the phases, and in particular their filler content in each polymer phase (distribution). Due to the successive mixing stages, not only is its energy consumption increased but the thermo-mechanical degradation of polymer chains may also be promoted. However, the absence of reactive ingredients permits to increase the mixing temperature to an elevated value that is lower than the level at which polymer thermal degradation or premature oxidation can occur.

In the conventional final mixing process the cold masterbatch, the curatives, and the reworks are fed into the mixer, where the masterbatch and reworks first have to be plasticized before the curatives are mixed in. As soon as a critical temperature limit is reached, the process has to be terminated to prevent scorch formation. This can cause insufficient homogeneity in curatives and lead to vulcanizates with an inhomogeneous distribution of crosslinks. To improve distribution of both the curatives and rework, additional two-roll mills are employed.

Mixing processes which blend ready-made rubber/carbon black batches are aimed at improving the level of distribution in the blend or preventing filler transfer. They are usually used when rubbers with highly different polarities and double bond content are blended with one another. The degree of filler dispersion and the homogeneity in the respective batches are related or of consequence to the properties of blends made from masterbatches.

6.4.2 Temperature, Torque, and Power Consumption

Useful information about filler incorporation, dispersion, and distribution is gained from control operations which measure stock temperature, torque, power consumption, ram displacement and, in the case of CB filled systems, electrical conductivity. A torque vs. time curve provides information about the mixing phases. Typical mixing diagrams are shown in Chapter 2.

In earlier publications it has been suggested that specific mixing energy is independent of machine design and/or size [4]. This assumes that all mixers are equally efficient in their use of supplied energy and, in addition, that the mixing process follows the same path for a given mix quality. The first assumption has to be proven in order to accept the latter. Although the specific mixing energy is mostly calculated from the energy input (integral of torque over mixing time) and not from a detailed energy balance, it can be stated that the "goodness" of different mixes at the same specific mixing energy is related to the main compound ingredients.

Special effort was made to monitor mixing of CB in rubber by electrical conductivity. One of the best solutions to perform reliable on-line measurements is described in [96]. Placing one ring electrode in the tip of a rotor flight and the counter electrode on the wall of the mixing chamber, a much higher number of signals can be obtained than by using one electrode in the rotor and the other in the wall [97]. At the beginning of the incorporation phase the conductivity is high due to non-incorporated pellet fragments and agglomerates (Fig. 6.9).

The wetting and encapsulation of the filler particles by isolating polymer reduces conductivity. The subsequent dispersion of the pellet fragments results in a higher density of conductive particles per unit volume. Consequently, a characteristic increase of electrical conductivity to a maximum, which corresponds to the maxima in torque and power consumption, can be noted. The distribution of filler particles leads to a decrease in electrical conductivity. In-house lab tests have shown that when the maximum macro-dispersion is attained, the variation of the conductivity signal reaches a minimum value. This behavior can be used to control

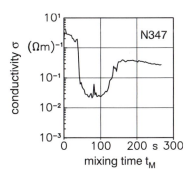

Figure 6.9 Conductivity versus mixing time for a masterbatch process

the mixing process. It has to be mentioned that the electrical conductivity reached at the end of the mixing process is not necessarily an equilibrium value. During storage of the mix its electrical conductivity increases due to re-agglomeration (flocculation) of the filler network (see the following).

6.4.3 Mixing Time and Rotor Speed

The majority of investigations reveal that "macro-dispersion" tends to reach a limiting value when mixing is prolonged. Combined action of rotor speed and mixing time leads to a plateau value of CB "macro-dispersion" that cannot be surpassed under the chosen mixing conditions (Fig. 6.10).

The plateau is a consequence of both the increasing temperature and the decreasing viscosity of the mix. As a result, shear stress is transmitted less efficiently and the adsorption of the polymer chains on the filler surface is decreasing. After cooling down, followed by a milling procedure (i.e., on a two-roll mill) the dispersion index in the plateau region can increase further. However, investigations of physical vulcanizate properties indicate that "micro-dispersion" is further developed during the mixing action in the plateau region. This is also supported by TEM investigations that confirmed agglomerate size distributions become narrower when compounding operations last longer [98]. By additional rigorous mixing of a system at the percolation threshold, the CB aggregates are separated from each other and percolate at higher loadings. In some publications a breakdown of carbon black aggregates during mixing is discussed as a supplementary dispersion mechanism. However, it is more likely that agglomerates are degraded to single

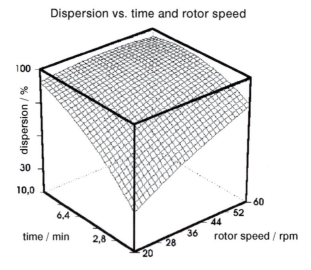

Figure 6.10 Influence of rotor speed and mixing time on CB "macro-dispersion"

aggregates. This interpretation is supported by TEM investigations of the solid content in agglomerates during mixing [98].

6.4.4 Cooling

By reducing the temperature of the cooling water, the incorporation phase is prolonged but the dispersion rate and dispersion index reach higher values (Fig. 6.11). Due to the higher viscosity, the polymer cannot be squeezed quickly into the voids of the agglomerates and thus the incorporation phase lasts longer [96]. The more important increase in dispersion rate and the final dispersion index can determine and influence the essential properties of the final products. TEM investigations confirmed that the improvement in "micro-dispersion" is the more important effect, realized by better cooling. Better cooling is obtained as a result of the larger surface/volume ratio of the mixers with intermeshing rotors compared to those with tangential rotors (see Chapter 2).

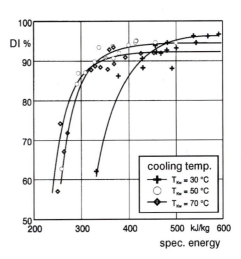

Figure 6.11 Influence of cooling on incorporation time and CB dispersion [96]

6.4.5 Alternative Dispersion Techniques

An enormous amount of work can be found in the literature on the issue of CB dispersion improvement. Because of the limitations cited, a compound can rarely – if ever – made economically that is completely free of undispersed filler content. This is due to the form in which the filler is supplied – as hard pellets – but also to the relatively large clearance of the compounding tools. Certain successful developments have been realized to attain relevant nanoscale dispersion in master batches.

By mixing polymer latex with filler slurry, then coagulating the mixture chemically and submitting the material to spray drying, a free flowing rubber-filler composite (RFC, Degussa) was produced. This material is suitable for both batch-wise processing in an internal mixer and continuous mixing on co-rotating twin screw extruders [99, 100]. With RFC, the filler dispersion can generally be improved relative to dry mixing [101]. However, the long mixing and coagulation time reduces productivity. In addition, for some polymer latexes, such as natural rubber (NR), there are some non-rubber substances, protein in particular, which can be adsorbed on the CB surface, hence interfering with polymer-filler interaction.

During the last decade, great effort has been made by Cabot Corporation to develop a unique continuous liquid phase mixing/coagulation process to produce a natural rubber-carbon black masterbatch [102] (called Cabot Elastomer Composite, CEC). The production process consists of CB slurry make-up, NR latex storage, mixing of both components in a special reactor and coagulation of the CB slurry and latex, dewatering of the coagulum, drying, finishing and packaging [103.] Under highly energetic and turbulent conditions, the mixing and coagulation of the polymer with the filler is completed mechanically at room temperature in less than 0.1 second, without the aid of chemicals. The dispersion obtained in this type of masterbatch is superior to conventional mixed batches made from the same compound components [104]. Starting from an initial high degree of dispersion within the CEC (without filler pellets), the mixing time can be substantially reduced and the quality of the mix is always higher than that of conventional batch mixes [104].

Recently, an efficient process for dispersing clay in rubber latex was presented. The dispersion work was carried out in an extensional and/or turbulent flow [105]. By means of "Continuous Dynamic Latex Compounding"(CDLC), working with aqueous slurry of clay and rubber latex, it proved possible to attain degrees of dispersion that mechanical compounding tools cannot achieve, even after repeated compounding cycles [106]. The CDLC process uses very short residence times.

6.5 Materials Influences on Filler Dispersion

As far as filler dispersion is concerned, the influence of the compound ingredients that support the disintegration of the filler agglomerates under hydrodynamic stress and facilitate micro dispersion is of similar importance as the influences of processing conditions. There are two groups of properties that have to be considered (see Section 6.2.3). The properties of the polymer facilitating wetting of the filler also lead to the formation of an interfacial layer through which the shear stress can be transmitted to the agglomerate. The other group includes properties of the filler particles that control the cohesive strength of agglomerates and provide the physical binding sites for an efficient wetting. The main contributions to the cohesive strength of the agglomerates depend on surface energy site distribution and functional groups as well as the inter-aggregate connections within the agglomerates. These parameters determine the rate of dispersion and the final yield of small agglomerates.

6.5.1 Influence of the Polymer

In a rubber melt, the chains are entangled and establish weak intermolecular interactions [107]. If filler particles are introduced into the melt, chain segments naturally tend to interact with the filler surface forming an interface. The strength of these interactions depends, in part, on the chemical nature of the polymer and on the surface activity of the filler. A necessary precondition for a potential interaction between polymer and fillers is that both components have corresponding interactive groups that result in a preferential adsorption of the chains on the filler surface [10]. Due to the interaction between polymer and filler, the polymer chains can be adsorbed either physically onto the surface or chemically. The result of the adsorption process is the formation of bound rubber and a rubber shell on the filler surface, both of which are related to the reduction of chain dynamics.

6.5.1.1 Adsorption from Solution and Melt

The adsorption of a polymer chain on the filler surface takes place by anchoring small sequences of chain segments (trains) on energetic sites of the solid surface. The chains between the anchored trains will tend to maintain their coiled shape, forming loops, which can be entangled with other free chains from the polymer matrix [108]. The changes in configurational entropy associated with the adsorption are compensated by the exothermic energy released by forming the favorable polymer-filler contacts (segment trains). However, it has to be taken into account that the segment trains are in dynamic balance on the solid surface (i.e., some segments are released from the surface, while at the same time others become bound to it). The chain segments can also continue to rearrange themselves on the solid surfaces – due to the influence of external forces on the solid surface, for example – and thus change their binding sites. Due to energetic considerations, chain segments will adsorb on energy sites and move on the surface as long as the strongest interaction sites are found. Such molecular slippage mechanisms are discussed in particular for stress softening of CB filled elastomers [109]. The influence of polymer properties on adsorption can be demonstrated best by "competitive" adsorption experiments in solution [40]. If polymers with different chain lengths (and narrow molecular weight distribution) and with identical chemical nature (e.g., polyisoprene) are placed in a CB suspension, it is observed that the chains are adsorbed in the sequence of decreasing chain length. This observation has been confirmed by the theoretical models that predict the number of contacts of a polymer chain on the filler surface to be proportional to the square root of the degree of polymerization (or chain length) [108].

$$v_P \sim P^{1/2} \tag{6.11}$$

where v_P is the average number of contacts with the surface established by a chain and P is the degree of polymerization.

It has further been determined that the chains are physically bound on the surface. In order to remove a chain, it would be necessary to release all contacts at the same time, this is rather unlikely to happen for long chains, unless additional energy (heat) or a good solvent enters into play. By heating the solution, physically adsorbed chains are quantitatively released from the filler [40].

The effect of the polarizability of chain segments was investigated by "competitive" adsorption of a high molecular weight polymer with less polarizable segments (i.e., polyisoprene) and a low molecular weight polymer with high polarizable segments (i.e., polystyrene). Despite its smaller chain length, the latter species are preferentially adsorbed on CB surfaces [110]. Preferential adsorption is based on the interactions of the phenyl groups of polystyrene with the aromatic species of the CB surface. This is proven by a shift of the H-out-of-plane vibration band of the phenyl group (at 699 cm^{-1}) [110]. This observation highlights the importance of the energetic contributions of chain segments to the adsorption process on the filler surface. By comparing the adsorption behavior of polymers with a similar degree of polymerization but different chemical constitution it is found that the share of a particular polymer in the polymer layer attached to the filler surface is always higher for polymers with a stronger interaction potential. Taking the solubility parameter δ as a measure for the average interaction potential, it can be shown that the polymer with the higher δ-parameter is preferentially absorbed on the filler surface. By increasing the thermodynamic similarity of the polymers (simlar δ-parameters), the partition ratio of the polymer species on the adsorbed layer becomes more equal, as shown in Table 6.1) [35]. Apparently the striking result that IR/1.2-BR with only a negligible difference in d-parameter of the polymers does not show an equal partition can be explained by taking into account the shorter chain length of 1,2-BR due to 1,2 insertion of butadiene units and implicitly a reduced number of contacts per contour length of the polymer.

The absorption of polymers from the melt confirms, on the one hand, the determinative influence of molecular weight but also the impact of the polymer's chemical composition. Bound-rubber is understood as the polymer portion in an uncured compound which cannot be extracted by a good solvent due to the adsorption of chains. The phenomenon is recognized as a characteristic feature of filler surface

Table 6.1 Partition Ratio in Competitive Adsorption Experiments on CB N 330 [35]

Polymer systems	$|\delta_1 - \delta_2|$ (J/cm³)$^{1/2}$	Partition ratio
PS/IR	3.31	100/0
PS/cis-1,4-BR	2.72	95/5
cis-1,4-BR/1,2-BR	0.65	75/25
cis-1,4-BR/IR	0.59	60/40
IR/1,2-BR	0.06	70/30

activity. It has been reported that bound-rubber content increases with the concentration of double-bonds, the amount of styrene units and functional groups in the rubber chains [111–113]. Analysis of the molecular weight distribution of the rubber extractable with solvents confirms that predominantly longer chains are contained in bound rubber [114]. With increasing compounding time, an increase in the bound-rubber content is found during dispersion and distribution, hence confirming the stronger wetting of the filler surface with rubber [115].

6.5.1.2 Influence of the Polymer on Filler Dispersion

The expectations from model adsorption experiments are confirmed by investigations on filler macro-dispersion. Under constant mixing conditions, high molecular weight NR provides a significant higher dispersion index than low molecular weight NR (Fig. 6.12a) [116]. The beneficial effects of the molecular weight is no longer present for high filler loadings (≥ 80 phr) and the dispersion index drops to

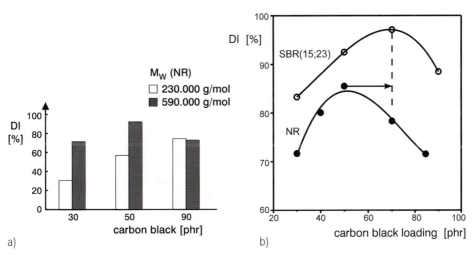

Figure 6.12 Influence of a) molecular weight and viscosity of the polymer and b) of the chemical nature of the polymer on "macro-dispersion" of CB

lower values due to a molecular weight breakdown (reduction in viscosity) under shear forces.

The influence of chain length and the potential interaction of chain segments can be confirmed by the results obtained from the comparison of CB dispersion in E-SBR and high molecular weight NR. By investigating the dispersion index under constant mixing conditions in high viscous NR and in low viscous SBR it was shown that the stronger interacting SBR systematically provided a higher dispersion index than the more viscous NR (Fig. 6.12 b). Interestingly, the peak values of the two curves are located at different values.

Among others, the "molecular friction" of chains [117] and chain stiffness play a role with regard to "macro-dispersion". This was demonstrated by using SBR grades with constant styrene content but with variable vinyl-content [112]. The examination of the optical roughness reveals a better "macro-dispersion" caused by a higher content of vinyl-units in SBR. The quite unexpected result can be explained by the reduction in chain flexibility – caused by vinyl-side groups – that contributes to the acceleration of the breakdown of pellets in the shear field.

Investigations on "micro-dispersion" performed by TEM demonstrate that the size distribution is dependent on the type of polymer: the better interacting polymer provides the narrower agglomerate size distribution [96, 98]. The influence on "micro-dispersion" is evident in the electrical percolation threshold. By keeping the type of filler constant, the electrical percolation threshold increases from low concentrations in saturated rubbers to high concentrations in unsaturated rubbers with higher polarizability (Fig. 6.13) [118].

In the case of saturated chain backbone with low polarizability, the attractive inter-aggregate interactions are much stronger compared to the weak polymer-filler interactions, leading to percolation threshold at lower concentrations. By increas-

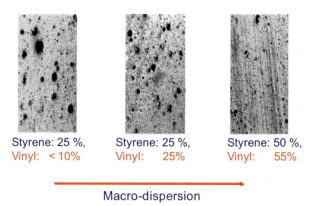

Figure 6.13 Correlation between the polymer solubility parameter and the electrical percolation threshold of CB N330

ing the unsaturation and especially by introducing efficiently interacting styrene groups into the chain, the filler network is formed by better dispersed (smaller) particles at higher CB concentrations. It was reported that introducing specific interacting functional groups into the polymer, the percolation threshold can effectively be pushed to higher concentrations [119, 120]. This example proves that under constant compounding conditions, the degree of micro-dispersion and the quality of the compound is shaped and influenced largely by the polymer.

6.5.2 Influence of the Filler Morphology and Surface Properties

The impact of filler characteristics on dispersion is essentially associated with the aggregate size, the aggregate shape and "structure", and most importantly to a certain degree to surface activity.

6.5.2.1 Influence of Surface Specific Area

It was proven in many studies that the bound rubber increases (at constant loading) with the surface area of the filler [121]. The result is certainly related to differences in interfacial area in the compounds between different grades of CBs. It can also relate to both the extent of the interface or to a stronger packed interface (more adsorbed chains per surface unit). From this observation it can be concluded that the surface specific area contributes to a better dispersion. However, a wealth of experimental expertise confirms that the dispersion of CB becomes more difficult when the aggregate size decreases. With low structure CB groups, an increase in the specific surface area from ca. 30 m^2/g to 140 m^2/g leads to an improvement in dispersion index (measured by optical surface roughness) from 75 to 85% only. Whereas with high structure CB groups, the same increase of the specific surface area leads to a decrease of the dispersion index from 95 to 90%. Traditionally, CBs with surface areas higher than 160 m^2/g and CDBP lower than 60 mL/100 g cannot be dispersed by dry mixing using the existing equipment; they are not considered rubber grades [14].

It was observed that the percolation threshold of CB filled compounds is shifted towards higher filler concentrations when the filler aggregate size increases [84]. The shift of the electrical percolation threshold is relevant, because it covers a wide range from ca. 20 to 70 phr. This result cannot be explained by the effect of particle size only. It reflects primarily the influence of the surface activity that decreases with increasing aggregate size (Fig. 6.14).

Taking into account recent results regarding the energy site distribution (see Section 6.2.3), it becomes clear that small filler particles with a higher frequency of high energy sites on the surface unit area will reach the percolation threshold at lower concentrations. As a matter of fact, if dispersion is facilitated by efficient

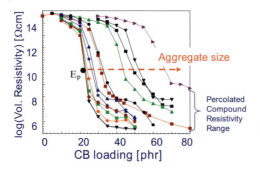

Figure 6.14 Electrical percolation threshold for CB as a function of the aggregate size

polymer-filler interactions and intensive mixing, the percolation threshold is shifted towards higher concentrations. Therefore, the filler volume fraction at the percolation threshold is a reliable criterion for evaluating "micro-dispersion".

6.5.2.2 Influence of Structure

As a consequence of the fractal nature of the aggregates and their relation to the connection number v_F, the dispersion index can be increased by employing high-structure blacks (HS-CB) with large DBP-numbers [36]. The fundamental interrelations between surface specific area, structure, and macro-dispersion were investigated for an oil extended SBR 1712 compound containing 80 ph CB (Fig. 6.15). The contour lines represent a constant dispersion index measured by the stylus roughness method (Section 6.4.1.3) [122].

It can be seen that the DBP-number has a determining effect on "macro-dispersion" and the surface area effects are somewhat varied. By using low-structure blacks (LS-CBs) it was observed that no satisfactory dispersion levels can be reached, especially at short mixing times, irrespective of their surface specific area. This disadvantage might be partly compensated by the shorter incorporation time of these fillers. The dispersion index increases slightly with increasing surface specific area. If intermediate DBP-levels are considered, the influence of the surface specific area is minimal. For HS-CBs the registered dispersion index is

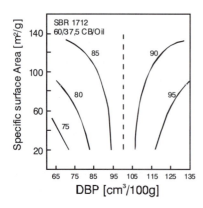

Figure 6.15 Macro-dispersion of CB as a function of the surface specific area and the DBP number of CB

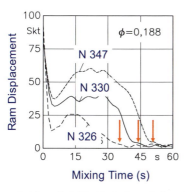 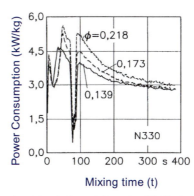

Figures 6.16 and 6.17 Evolution of electrical conductivity during mixing of a CB filled compound (50 phr CB N347/SBR) [96]

always on a high level [122]. Among the CBs with similar surface area it was found that HS-CBs demonstrate higher bound-rubber content. This was related to (i) segment adsorption [122], (ii) less ordered graphite micro-crystallites, and (iii) easier aggregate breakdown during mixing [123].

The varying cohesive strength of the filler agglomerates has an impact on the course of the compounding process and becomes evident in the time dependency of process variables, such as ram displacement, torque power consumption, and electrical conductivity. The curve of ram displacement and specific power consumption is plotted as an example (see Figs. 6.16 and 6.17). It is particularly noticeable that in the case of low-structure CB, the ram comes to rest relatively early, corresponding to a short incorporation stage, and the integral power consumption is low on the whole, indicating insufficient dispersion work. In the case of HS-CB, both assessment criteria are more pronounced and differ more strikingly in terms of time (see Fig. 6.18, page 210).

6.5.2.3 Influence of Filler Surface Activity

There are many observations indicating the importance of filler surface activity in polymer reinforcement. It is reasonable to assume that on the one hand surface activity can promote the attachment of polymer chains to particle surfaces, leading to polymer-filler interfaces and on the other hand, the inter-aggregate connections are strengthened by surface activity. It was already shown that dispersion is impaired for fillers with high surface activity. In order to overcome the detrimental effects on cohesive forces within agglomerates, higher mixing energies and changes in the mixing procedure are required. According to Eq. 6.6, a good balance between polymer-filler and filler-filler interaction has to be established in order to predict the effects of these complex influences.

The existence of physical adsorption resulting in bound-rubber and rubber shell formation has been demonstrated in a TEM study of SBR samples, in which the sol was extracted nearly completely. The small amount of polymer remains on discrete patches of the surface. [124].

If the bound rubber content in a series of CB filled compounds is normalized with the interfacial area, it becomes obvious that the amount of polymer adsorbed per unit filler surface decreases with decreasing particle size of the filler. The striking but experimentally reproducible result is in contrast to the observations made in surface energy measurements, where more energy sites per unit surface are detected for CBs with smaller aggregate size [63, 65]. The results were interpreted in terms of inter-aggregate multi-attachment of rubber chains and degree of dispersion of CB. They clearly indicate that under identical compounding conditions the theoretically available interface of the filler cannot be established, due to inter-aggregate interactions, which increase with decreasing particle size. In a first approximation it can be stated that the filler dispersion decreases with increasing surface specific surface [121]. However, the loading dependence of the bound rubber per unit surface of CB shows also lower values for the higher loadings, which is related to the inter-aggregate multi-attachment.

When graphitized CBs (N234) are compounded into E-SBR, the dispersion level decreases with increasing treatment temperature. Because over the range of temperatures used the morphologies of the CB cannot be changed, the variation of dispersion of CD after heat treatment can only be interpreted in terms of their surface characteristics. The heat treatment results in a change of energy site distribution. That is reflected in the properties related to adsorption:

- the surface specific area determined by nitrogen adsorption is lower for graphitized CBs than for corresponding ASTM grades [125],
- the bound rubber content determined for graphitized CBs is extremely low [126].

However, in graphitized CBs even the reduced filler-filler interaction exceeds the polymer-filler interaction. Thus, the percolation threshold of graphitized CBs occurs at considerable lower concentrations ($\leq$ 15 phr) than for ASTM grades [84].

The extent to which the localized energy site or the surface roughness is responsible for dispersion and reinforcement is a subject still not fully understood. In this respect, it correlates to some of the controversial issues that still remain unresolved. While the energy site distribution is strongly dependent on aggregate size, the surface roughness expressed by the surface fractal dimensions d_s of ASTM-blacks increases only slightly with surface specific area (d_s = 2.59 – 2.63). Every change in CB manufacturing that could lead to a product with a higher amount of amorphous carbon on the surface and less graphitic micro-crystallites will be a step towards better dispersible CB [125].

Compared to carbon black, silica can form agglomerates with high cohesive strength caused by hydrogen bonding of silanole groups [9]. Therefore, in a hydrocarbon polymer the polymer-silica interaction, which is determined by the dispersive component of filler surface energy, is lower than the inter-particle interaction, which is supported by the polar silanol groups. Consequently, the poor compatibility with the polymer and the strong filler–filler interaction due to the high surface activity lead to a stronger filler network and a percolation threshold at lower concentrations. Changing the surface activity of silica by treatment with mono- or bifunctional silanes, the filler dispersion in polydienes can significantly improve the substitution of the most abundant part of the silanole groups by non-polar alkyl chains, and this in turn decreases significantly the interaction potential [127].

From the thermodynamic point of view, there are a few approaches available to improve micro-dispersion of the filler [14]:

1. Reduce the difference in surface characteristics, especially in surface energy, between polymer and filler by filler surface modification and/or polymer modification.
2. Increase interaction, affinity, or/and compatibility between polymer and filler surface by using chemical or physical coupling agents.
3. Use fillers with hybrid surface characteristics.

The surface activity of carbon blacks can be changed by changes in process technology able to reduce the size and amount of low energy graphitic micro-crystallites while increasing the share of adsorption sites with higher energy [128]. Other possibilities include a controlled surface treatment with chemicals or by plasma [129, 130]. The aim of such post-treatment is to explore the routes to improve surface activity by small changes of surface chemistry, to reduce the concentration of the very high energy sites, and to weaken the inter-aggregate interactions.

Recently, several approaches for CB modification have been patented [131 – 133]. A chemical approach is based on the decomposition of a diazonium compound that results in the attachment of aromatic compounds onto the surface [136]. The surface chemistry can then be tailored meeting the requirements of various applications. Treatment of the surface can reduce the inter-aggregate interaction, resulting in better filler dispersion. Chemical reactions with small quantities of zinc-dithiocarbomate or peroxides have a similar effect [129]. The treatment of fluffy CB with atmospheric N_2-plasma in a flow reactor can result in better dispersion and homogenization of compounds [130]. A promising approach to improve CB properties related to micro-dispersion was the development of carbon-silica dual phase fillers [134] based on a co-fuming process in the CB reactor [135].

6.5.2.4 Dispersion Kinetics

At shorter mixing times, as are usual in practice, significant differences in the rate of incorporation and dispersion can be observed, generated by the nature of the polymer and the type of the filler. Polymers with a high molecular weight (or viscosity) require more time to penetrate the agglomerate voids during the incor-

Table 6.2 Comparison of DI of CBs as a function of surface specific area DBP-number

CB	CTAB (m²/g)	DBP Ml/100 g	DI (%)	Distribution spread
1	119	125	94	0.15
2	136	119	70	0.15
3	162	119	49	0.23

poration phase and to incorporate the filler. The somewhat longer incorporation phase is offset by a shorter dispersion stage, which derives from higher shear stress and more efficient wetting. As mentioned, the rigidity of the polymers employed and the molecular friction coefficient have a positive effect on the breakdown of pellets and the decomposition of larger agglomerates [112].

In the case of fillers, the compounding operation is promoted mainly by agglomerate structure. Although the incorporation time is somewhat longer for high-structure fillers, the dispersion stage is shorter because of the reduced cohesive strength of the filler agglomerates. A comparison of the dispersion index obtained under a constant set of processing conditions and short mixing times underlines the importance of agglomerate void volume versus surface specific area.

The dispersion process follows a first order time dependence. The rate constant for CB dispersion can be determined from the amount of pellet fragments and large agglomerates present in the mix (undispersed CB) as a function of mixing time. Even with short compounding times, a higher dispersion index (DI) is expected in the case of high-structure CB. Because of the faster pellet breakdown and the larger agglomerates in the case of HS-CBs, an ever more pronounced difference compared to the LS-CBs is also achieved with increasing compounding time. This difference is immediately reflected in compound quality. The example in Fig. 6.18 shows that

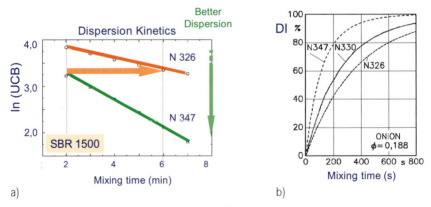

Figure 6.18 Dispersion kinetics for a) LS-CB (N326) and HS-CB (N347) and b) modellized according to the "onion skinning" model [96]

the dispersion stage for LS-CB would have to be considerably extended in order to achieve the same degree of dispersion as for compounds with HS-CB.

Mathematical modeling of macro-dispersion in accordance with the "onion skinning" model reflects what was learned rather well experimentally [96]. More in-depth research into these mechanisms is important and is the overriding factor in the development of high-performance elastomer components.

With NR/E-SBR blends it has been established that there is a reduction in the carbon black incorporation phase when NR is added. If NR forms the continuous phase, the dispersion level of the blend achieved approaches that of pure NR. The dispersion rate reaches its highest value in the phase inversion stage. As with NR/E-SBR, a reduction of the incorporation time compared to that for pure components has also been observed for NR/L-SBR blends. However, unlike the previous system, the presence of L-SBR results in a reduced dispersion rate in the phase inversion stage [136].

6.5.2.5 Filler Re-Agglomeration

A certain amount of energy input is needed during mixing to break down the agglomerates and to disperse the aggregates in a polymer matrix. It has been demonstrated, however, that when the filler is well dispersed in the polymer, the aggregates tend to re-agglomerate during storage and vulcanization of the uncured compound, especially at high loading [137, 138]. The process is a natural consequence of inter-aggregate interactions, which comes into play when the mean spacing of the dispersed particles falls below a critical value. As the surface energy of hydrocarbon rubbers is generally lower than the one for CB or silica, the driving force behind the re-agglomeration is the reduction in interfacial energy and in the contact surface.

It should be pointed out that the re-agglomerated black in the polymer is different from undispersed black. In the former case, the rubber can be trapped in the agglomerates, but for the latter, no polymer penetrates into the agglomerates. Because of the flow rates prevailing in the compounder, this operation can hardly be documented during the compounding process. However, as soon as the compound is stored and re-agglomeration is no longer prohibited by shearing fields, an increase in the compound's electronic conductivity, viscosity, or elasticity modulus can be observed as a consequence of this process. The study of re-agglomeration rate shows that the process is promoted by low rubber viscosity or elevated temperatures. Fillers with a higher surface activity have a stronger tendency to re-agglomerate than less energetic fillers.

Because of their polar surface, non-silanized silica tends to undergo a particularly fast and pronounced re-agglomeration. Surface silanization greatly weakens this tendency in nonpolar hydrocarbon rubber [139]. The re-agglomeration of fillers imparts a significant effect on the properties of filled rubber, especially rheological properties of the uncured compounds and viscoelastic properties of the vulcanizates.

6.5.3 Influence of Oil on Filler Dispersion

In compound manufacturing it is necessary to use plasticizing process oils, because highly filled rubber compounds can attain high viscosity (or even a flow limit), rendering processing more problematic. To ensure good processability, different amounts of mineral oil are proportioned. On the basis of their chemical composition as hydrocarbons, these compound ingredients are similar to nonpolar rubbers. In the compound, the mineral oil therefore has a tendency to compete with the rubber constituents for the energetic sites on the filler surface. Because of their low molecular weight, the mineral oil molecules are more mobile than long rubber chains, so that oil constituents can be adsorbed on the filler surface in the first phase of the dispersion process, thereby partially or completely hindering wetting by long rubber chains. As a consequence, the required hydrodynamic stress cannot be applied in the dispersion phase, resulting in a marked impairment of macro-dispersion. Figure 6.19 shows the effects of simultaneously mixing filler and oil [140].

If the oil is mixed after the filler has been incorporated and rubber wetting has already occurred, high dispersion index values are attained with low mixing energy, even for high admixtures of oil. Differences related to the quantity of mixed in oil are nonetheless evident. It should also be noted that the influence of the mixing in of oil is less pronounced in the case of large-particulate carbon black (low surface specific area) than it is in the case of fine-particle carbon black types. In actual practice, the admixture of carbon black and oil is sequentially phased, in

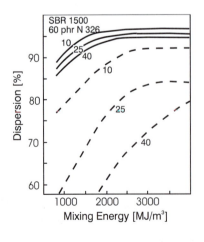

Figure 6.19 Influence of oil incorporation on macro-dispersion of CB

smaller portions. Resins and processing agents display the same "competition" for energy sites on filler surfaces. If larger quantities are mixed in, it is advisable to incorporate and disperse a portion of the filler beforehand.

6.6 Effects of Filler Dispersion on Material Behavior

A filled elastomer can be regarded as a composite characterized by a continuous "rubbery" phase and a particulate "rigid" dispersed phase. The continuous phase is responsible for the "rubbery" behavior of the composite, while the filler plays a specific role in changing various properties. Thus the mixing in of particulate fillers to a rubber compound alters the rheological behavior of the compound, with the degree of change in the characteristics determined chiefly by filler type and content and by the dispersion and distribution attained during mixing.

By crosslinking filled rubber compounds, a wide range of elastomers are obtained that have captured a fixed and indispensable place in the spectrum of modern composites, especially thanks to their dynamic-mechanical and ultimate properties. When such bodies are subjected to periodic oscillatory stresses – due to sinusoidal strains – the stress and the strain are out of phase (phase angle δ). In addition, when these materials are strained or stored, part of the energy input is dissipated as heat. Both the heat dissipation and the phase angle are characteristics of viscoelastic materials [107, 117].

6.6.1 Effects on Rheological Properties

The presence of the filler is known to change the rheological properties of the compounds, resulting not only in highly non-Newtonian flow at low shear rates, but also in comparatively high viscosity. The increase of viscosity can be attributed to several factors:

- the hydrodynamic effect of the filler loading reduces the volume fraction of the rubber and causes shear amplification in the matrix [141, 142],
- the "structure" or the anisometry of the filler increases flow resistance [142],
- the occlusion of rubber within the agglomerates increases the effective hydrodynamic volume of the filler [25],
- the adsorption of rubber on the filler surface [37].

Several authors have suggested a number of relationships between viscosity and filler loading (or volume fraction). The most successful approaches go back to Einstein's viscosity theory. They consider the hydrodynamic effect and the mutual disturbance caused by the increasing volume fraction of spherical particles in the polymer matrix [141, 142].

$$\eta = \eta_0 (1 + 2.5 \phi + 14.1 \phi^2) \tag{6.12}$$

where η and η_0 are the viscosity of unfilled and filled polymers, respectively, ϕ is the volume fraction of the dispersed filler in the medium.

It is to be noted that Eq. 6.12 refers to the filler volume fraction alone and not to the particle anisometry or the size of the particles. In analogy to short fibers embedded into a polymer melt, ellipsoidal or branched aggregates should also convey a higher resistance to shear flow. It can be assumed, though, that changes in the agglomerate size and shape will affect the composite viscosity. In addition, the portion of the polymer that is entering the void volume of agglomerates (occluded rubber) can cause an apparent shift in rubber/filler concentration, hence resulting in a higher viscosity of the mix. When structured CBs are dispersed in rubber, the rubber portion filling the internal void of the CB aggregates (or the rubber portion located within the irregular contours of the aggregates) is unable to participate fully in the macro-deformation of the filled system. The partial immobilization of the "occluded" rubber makes it behave as if it were part of the filler rather than the polymer matrix. Due to this phenomenon, the effective volume of the filler, with regard to the stress-strain behavior of the filled system, is increased considerably. The partial occlusion of polymer in the agglomerate voids changes the hydrodynamic relevant polymer/filler ratio. The more rubber is occluded in the agglomerates voids, the more the system behaves like a system with a higher filler content.

Consequently, HS-CBs always lead to higher compound viscosity. Empirical correlations between Mooney viscosity and structure measured by compressed DBP absorption can be established.

The adsorbed rubber on the filler surface results in complex effects [37, 143]. On the one hand, the rubber shell formed during mixing increases the hydrodynamic effective filler volume fraction. As large chains are adsorbed preferentially [29], the adsorption process may partly explain the increase in viscosity. On the other hand, the formation of filler networks is another reason for high viscosity of the filled compound and is responsible for influencing its non-Newtonian behavior.

Consequently, it was proposed to replace the filler volume fraction ϕ by an effective filler volume fraction ϕ_{eff} that empirically considers these complex influences. Substituting ϕ by ϕ_{eff}, it follows [144, 145]:

$$\eta = \eta_0 (1 + 2.5 \phi_{eff} + 14.1 \phi_{eff}^2) \tag{6.13}$$

The effective volume fraction ϕ_{eff} is always larger than ϕ by an amount related to, e.g., rubber shell or the occluded rubber. The effective filler volume fraction in a rubber mix was estimated from the void space determined by crushed DBP from the endpoint of DBP absorption [146, 147] and the equivalent spheres of aggregates (assuming the spheres to be packed at random).

Due to the fact that filler agglomerates are strongly degraded during the dispersion phase and separated from each other in the distribution phase, the degree of dispersion and distributiton co-influences the compound viscosity. Monitoring the effect of mixing time and rotor speed on the Mooney viscosity of compounds, measured after a storage period following mixing, a pronounced decrease in Mooney viscosity as a function of mixing time and rotor speed was observed. Accordingly, compound viscosity is expected to be reduced when employing a higher rotor speed or longer compounding times. With increasing compounding time, agglomerates are broken down with the corresponding hydrodynamic effects and the dispersed particles are separated from one another by a rubber layer. The Mooney viscosity response surface, shown in Fig. 6.20, is more or less inversely proportional to the response surface for the dispersion index (see Fig. 6.10). During a storage period the re-agglomeration of the filler (flocculation) is conveyed to the macroscopic scale by an increase in viscosity as well as in the storage modulus of the mix [139].

With two model fillers, CB and silica, having comparable surface areas and structures, it was demonstrated that the increase in Mooney viscosity after mixing was more pronounced and rapid in silica compounds due to more efficient inter-aggregate interactions promoting re-agglomeration [148].

In rubber processing, the elastic response of filled compounds is reflected in terms of die swell and the appearance of rubber extrudates. This phenomenon is associated with the elastic recovery caused by the incomplete release of long-chain molecules oriented in the shear field of the extruder die. Active fillers can reduce the elastic contribution and decrease the effective relaxation time of the mix. Thus, die swell is generally reduced by filler loading and become less pronounced when the "structure" of the filler increases [149]. The die swell of filled compounds has been shown to correlate with the filler structure [150]. The observation that die

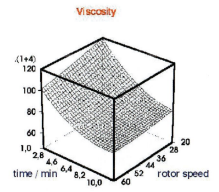

Figure 6.20 Influence of filler dispersion on the Mooney viscosity of filled rubber mixes

swell is lower than expected from the corrections for occluded rubber can be attributed to certain effects of dispersion. These can cause changes in the "in rubber" morphology of the filler agglomerates.

6.6.2 Effects on Dynamical-Mechanical Properties

Dynamic mechanical properties of filler reinforced rubbers have been studied and reviewed intensively [151, 152]. It should be cautioned that the dynamic mechanical properties of filled rubbers are well defined only in small amplitude oscillatory deformations. Thus, the storage modulus *(G')* and the loss modulus *(G")* can be used to define the behavior of the compounds. It was stated that at low strain (< 6-8%) the filler is the main contributor to reinforcement through the filler network. This network is sensitive to strain, diminishing the overall stiffness of the composite (*G'* decreases), whereas some of the network energy is dissipated as heat (*G"* passing through a maximum). It is remarkable that after the strain is removed, the network is partially reformed. There are reasons to assume that the stiffness of the unstrained composite is proportional to the stiffness of the filler network [153]. Because of the filler network formation it becomes obvious that the state of dispersion and distribution reached is particularly important to low-strain dynamic properties.

6.6.2.1 Influence of Loading

Any kind of nanoscale, well dispersed filler would increase G' and $G"$ of a crosslinked compound in the T_g-regime. The loss modulus $G"$ increases proportionally to the filler volume fraction [127, 154]. To move from the glassy to the amorphous state, an additional amount of energy is required to increase the chain mobility inside the adsorbed rubber shell. For a constant composition and dispersion, the slope of the linear relationship describes the effect of interface on energy dissipation in the T_g-regime. The plot takes into account neither the shifts in T_g brought about by the filler nor the inherent widening of the damping maxima:

$$G"_{(F)} = G"_{(P)} (1 + \alpha \phi) \tag{6.14}$$

The slope α is designated as the "interaction" factor. It determines system-specific interactions, such as polymer-filler interaction and filler dispersion [155]. It was observed that the interaction factor α is sensitive to any change in polymer-filler affinity and increases when filler dispersion is improved, e.g., by filler surface modification. The relationship is the base for the calibrations needed to distinguish the filler partition in rubber blends from the dynamic mechanical properties (see Section 6.7).

The storage modulus (G') and loss modulus (G") in the rubbery plateau (above T_g) exhibit at small volume fractions a typical increase, which is attributed to hydrodynamic reinforcing effects, and a function of the effective filler volume fraction ϕ_{eff} [145]:

$$G_{\gamma\to 0} = G_0 (1 + 2.5\,\phi_{eff} + 14.1\,\phi_{eff}^2 \ldots) \qquad (6.15)$$

With the concept of "occluded" rubber the increase in G' can be described [146]. As a direct consequence, a desired G' value can be obtained by using a HS-CB at lower loading or a LS-CB at higher loading (structure-concentration equivalence principle) [147].

It was observed that a systematic increase of the filler loading will result in a non-linear increase of $G'_{\gamma\to 0}$. Above the percolation threshold, the storage modulus increases exponentially as a function of the filler volume fraction [157]:

$$G'_0 \sim \phi^\alpha \qquad (6.16)$$

In percolated systems submitted to very small strain amplitudes, the macroscopic stress is no longer transmitted through the rubber matrix alone, but merely through the filler network [45, 156, 157].

Assuming a similar structure of the filler network [45], the exponent in Eq. 6.16 is predicted to be 3.5. Experimental results do confirm this value in some cases. For many systems it was found that the exponent is less than 3.5 [106]. However, as shown before, the percolation threshold depends on the type of the filler and the polymer [158]. Because the percolation threshold is a function of polymer-filler interaction, comparisons of G' for different compounds should be treated with caution. Taking into account that better polymer-filler interaction and filler micro-dispersion is shifting the percolation threshold – above of which Eq. 6.16 comes into play – towards higher filler loadings, it becomes clear that at a given filler loading the systems with the lower percolation threshold will exhibit higher G' values than systems with high micro-dispersion and high percolation threshold [106, 158].

6.6.2.2 Strain Dependency

While the storage modulus of the unfilled compound does not change upon increasing strain amplitude, it decreases significantly for the filled rubber [159, 160]. This non-linear behavior is clearly described by the strain dependence of the complex modulus on the strain amplitude (γ):

$$G^*_F = G'_F(\gamma) + iG''_F(\gamma) \qquad (6.17)$$

The strain amplitude seems to be the most important testing parameter when studying the role of fillers in rubber compounds subjected to low frequency deforma-

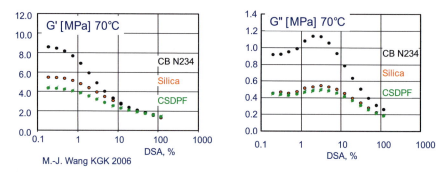

Figure 6.21 Strain amplitude dependency of G' and G'' as a function of filler surface activity [200]

tion. For compounds with a filler content above the percolation threshold, the storage modulus G' decreases in a characteristic manner and the loss modulus G'' passes through a maximum when the sample is submitted to oscillatory deformations with increasing strain amplitudes (Fig. 6.21).

In the limiting case of high strain amplitudes, a constant G' value is asymptotically approached. This non-linear behavior was intensively examined [159, 160] and is known as the "Payne" effect. The interpretation of this effect relies on mechanically unstable filler-filler interactions. When local deformation becomes large, the cohesive connections within the cluster network are continuously released and the entire filler network is gradually decomposed into sub-networks. The decrease from G'_0 to the limiting value G'_∞ was modeled in several theoretical approaches:

- the dynamic network model [161]
- the adhesion model [162]
- the cluster-cluster aggregation model [45]

The low strain limit G'_0 primarily represents the mechanical response of the filler network and additional contributions by mechanically stable filler-filler interactions (occluded rubber) and hydrodynamic reinforcing effects, as well as by the crosslink density of the rubber phase. The difference between the two limits, $G'_0 - G'_\infty$, is attributed to the mechanical effect of the filler network.

The influence of the filler dispersion becomes evident primarily from the decrease in G'_0 if mixing time is lengthened or rotor speed is increased. A real-life criterion for assessing dispersion can be derived from the decrease in G'_0 within a defined deformation range. Because improvement in filler dispersion is related to a smaller value of G'_0 and a flatter $G'(\gamma)$ function, it can be expected that the $\Delta G'$ measured in the selected range of strain amplitude becomes smaller when the dispersion increases [163]. By improving dispersion due to more efficient mixing or controlled changes in filler surface activity, the maximum of the function $G''(\gamma)$ decreases.

6.6.2.3 Effect of Filler Surface Modification

Any change in filler surface activity that reduces the cohesive forces in agglomerates will be reflected in (i) a reduction of $G'_0(\gamma)$ and $G''_{max}(\gamma)$ and (ii) a shift of the percolation threshold towards higher filler volume fractions. This was clearly demonstrated for silica filled BR compounds, were the filler surface was modified by monofunctional silanes [155]. It was observed that the interaction factor α and the dispersion index increases with the alkyl chain length and the surface concentration of the silane on the silica. Caused by reduced filler-filler interactions and a better dispersion (higher percolation threshold), $G'_0(\gamma)$ decreases as a function of the degree of silanization and the alkyl chain length of the silane. Similar effects have been reported for carbon-silica dual fillers that show superior dispersion compared to ASTM blacks and silanized silica [14]. It has been shown that dispersion can also be improved by introducing functional groups into the polymer. The dispersion of precipitated silica in SBR containing 7 mol% epoxy-groups leads to a significant decrease of $G'_0(\gamma)$ and G''_{max} compared to the original SBR [164, 165]. Such changes are targeted to assure better traction and skid resistance in tire applications.

6.6.3 Effect on Ultimate Properties

The incorporation of reinforcing fillers changes the fracture behavior of rubber, leading to increasing tensile strength and dynamic cut growth resistance. Part of the mechanical energy supplied for crack propagation dissipates in the rubber, depending on the rate of tear propagation and temperature. The lower the temperature, the higher the contribution of viscoelastic energy loss mechanisms. The magnitude of the viscoelastic contributions to stresses depends on the rate of deformation. The presence of particulate fillers increases the hysteresis, because of several dissipation mechanisms, such as de-bonding within the interface and breakdown of filler structure. A relationship between energy dissipation on stretching and the energy density required to break was presented in [166]. The role of viscoelastic processes in steady-state tearing conditions was investigated considering the relationship between tearing energy and loss modulus [167]. It is generally accepted that cracks are initiated at an inherent flow, i.e., inclusions, microvoids, network inhomogeneities, and large filler agglomerates or pellet fragments [168].

Tensile Strength

Although the mechanism of tensile failure of elastomers has not been fully understood, it can be regarded as catastrophic tearing by growth of cracks initiated by accidental defects, such as large fragments of pellets, micro-wetting, or cavitations from the filler surface. When a vulcanizate undergoes stress, the local stress at the tip of the flow is magnified. Once the local stress at the tip reaches a critical level, which depends on the size of the flow and the rupture energy, cracks will be initiated and new surfaces will be created. It was theoretically predicted that if the radius of a microvoid is less than 0.1 µm, its surface energy will exert an additional restrain on expansion [168]. On the other hand, in case of poor polymer-filler interaction a de-bonding of the rubber shell or de-wetting will take place which can trigger crack initiation [169]. Obviously, in such systems the frequency of less dispersed filler particles and pellet fragments is higher.

At the same loading, fillers with small aggregate size and high surface activity will represent a high resistance against crack propagation. Since the crack cannot occur through filler particles, it has to move around the filler particle, thus requiring more energy. The effect is more pronounced for fillers with high surface activity and pronounced rubber shell. Since nano-scale fillers with large surface specific areas and high surface activity are conducive to processes that reduce crack initiation when well dispersed, it is not surprising that tensile strength increases with increasing surface specific area and surface activity, while the aggregate shape is less important. The degree of filler loading has the same effect as the surface specific area. Tensile strength increases up to a certain level and then decreases at higher filler concentrations. The concentration at which tensile strength reaches its maximum value depends on the type of filler and the nature of the rubber used in the compound. The maximum is attained at lower concentrations for fillers with small aggregate size and high surface activity. It corresponds roughly to the concentration at which the dispersion index also reaches a maximum value (see Section 6.6.1.2). The result clearly underlines the influence of filler dispersion on tensile strength. At high loadings the amount of pellet fragments or large agglomerates can lead to crack initiating flows. Fractographic observations showing that the flows size initiating tearing increases with filler concentrations support these observations [170]. It was shown that tensile strength is related to the degree of macro-dispersion. Any increase in mixing time and rotor speed during mixing leads to an improvement of tensile strength in the vulcanizate. However, as the mixing process is prolonged and an increasing rotor speed is employed, a similar plateau level for tensile strength is reached as it was observed for macro-dispersion (Fig. 6.22) [171].

A delay of crack initiation and an increase of tensile strength are achieved by using fillers with high surface area and surface activity. Assuming a good dispersion, the strongly attached rubber shell and the small radius of filler particles can prevent premature de-wetting of the polymer [172]. This is consistent with TEM investigations or cavitations in CB filled SBR compounds [173].

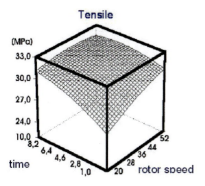

Figure 6.22 Influence of filler dispersion on tensile strength of SBR/CB N330 (50 phr) compounds

Dynamic Crack Growth

When dynamic crack propagation is considered, the above presented approach of control of macro-dispersion is no longer valid. It has been observed that in case of high strength, the tear deviates from a straight path to a tortuous one ("stick-slip" or "knotty" tearing) that indicates the role of both filler particles distribution and rubber-filler interface in repeatedly changing the direction of cut growth. When a long strip of rubber is subjected to cyclic extension, the crack behavior vs. tearing energy can be expressed as follows [174]:

$$\frac{dc}{dn} = BG^\beta \tag{6.18}$$

where c is the crack length, n is the number of cycles, G is the tearing energy and B and β are constants.

With regard to the effect of filler on dynamic cut growth flow, systematic investigations under controlled conditions have been carried out. The presence of fillers plays an important role, including the effects of poor dispersion, which affects primarily the effective initial flow size, and the strain energy [175]. Interestingly, it was observed that the dynamic cut growth resistance of CB filled SBR samples does not have limiting values when the mixing energy input is increased. Samples obtained from mixes prepared at gradually increased mixing time and rotor speed do not exhibit plateau values for dynamic cut growth resistance but instead show a pronounced tendency for steadily increasing values (Fig. 6.23). The result can be attributed to improved micro-dispersion, while the optically or mechanically monitored macro-dispersion is not significantly changed (see Fig. 6.10). During prolonged mixing action, as the concentration of small agglomerates and aggregates increases, the effective deviation of the cut path and the phase bonding increase. These are key parameters for a higher cut growth resistance.

Dynamic crack propagation experiments performed on a "Tear Analyzer" under repeated pulsed straining demonstrate the wide span of properties brought about by improving the state of filler dispersion. Evaluations of the crack propagation

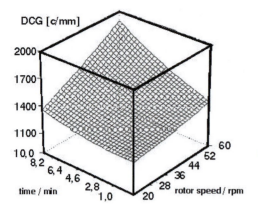

Figure 6.23 Influence of filler dispersion on dynamic cut growth resistance of SBR/CB N330 (phr) compounds

rate within the initial steady state portion of the curve – based on Eq. 6.16 – provide input data for fracture mechanical calculations. The beneficial effects of microdispersion on dynamic crack growth resistance were proven by comparing vulcanizates with identical composition made from rubber filler composites and from conventional SBR compounds. A fourfold (30 % strain) to tenfold (20 % strain) lower crack propagation rate was observed for the vulcanizates. The effects were even more pronounced at lower strains [101].

■ 6.7 Filler Distribution in Polymer Blends

High molecular weight polymers usually form phase separated blends with domains having similar physical properties as the original polymers. This beneficial situation offers the possibility to compound materials which combine conflicting properties. The use of blends remains at a high level throughout the rubber industry [34, 176, 177]. Major reasons pertain to reduced compound costs, simplifications in manufacturing and enhanced final product performance. For practical applications of reinforced rubber blends, knowledge about phase morphology and filler partition within the polymer phases is of major interest. In particular, this topic is important regarding dynamic properties, hysteresis, damping, tensile strength, abrasion, and wear [176]. This mainly involves the high frequency property regime that is closely related to the glass transition of rubber blends and the ultimate properties related to the materials life time. Due to the fact that usually rubbers are not miscible with each other in a thermodynamic sense, the blends obtained by mechanical mixing of rubbers contain discrete domains [177]. Fortunately, miscibility is not a requirement for most of the rubber applications. Homogeneity at a fairly fine level is necessary for optimum performance. More than the homogeneity of unfilled rubber blends, the filler phase distribution plays a significant role. It is important to ascertain the extent to which filler is partitioned to one of the other polymer phases. Also, the transfer of the filler from one polymer phase to the other during mixing is an issue determining the final properties of an elastomer. Miscible rubber blends might have the advantage of greater mechanical integrity because of the absence of polymer-polymer interfaces. However, there are only a few examples for miscible rubber blends.

6.7.1 Compatibility of Rubbers

Miscibility of polymers is a matter of both the molecular weight and the intermolecular interactions of chain segments. The former is basically responsible for the very low mixing entropy. By increasing the molecular weight, the chance to achieve miscibility becomes smaller. The latter is related to the chemical nature of the chain segments that determines the degree of interactions. Generally speaking, the similarity of the interaction potential of the blend constituents guarantees a good compatibility and in a few cases miscibility. The complete miscibility of polymers requires that the free energy of mixing be negative, which can be achieved by an exothermic heat of mixing or by large entropy of mixing [178]. Therefore, most rubber blends are phase separated, because the heat of mixing is endothermic and the entropy of mixing is very small due to the high molecular weight [34, 179].

Compatibility is very often used synonymously with miscibility. In materials technology, compatibility is a more general term with a wider diversity of meanings than miscibility, which is consistently defined by thermodynamic criteria. In a strict technological sense, compatibility is often used to describe whether a desired result occurs when two polymers are blended. A consistent description of polymer compatibility is achieved by employing the δ parameter concept developed by Hildebrand [30]. Basically, the compatibility between two unlike polymer increases when the difference in the δ parameter decreases. From theoretical considerations it was shown that $\Delta\delta$ is proportional to the interfacial tension.

$$\gamma \sim |\delta_1 - \delta_2| \tag{6.19}$$

Miscibility occurs for polymers having similar δ parameters so that the interfacial tension approaches zero ($\gamma \to 0$). Phase separation will be observed for rubbers whose δ parameters are different. The blend morphology is primarily governed by the interfacial tension and the processing conditions. The morphology of blends reveals a proportionality between the interfacial tension and the size of dispersed polymer domains [35] that can be detected by microscopic methods (i.e., optical microscopy, TEM, and AFM) [180–184]. In accordance to theoretical predictions it is found that the dispersion of polymer domains can be promoted by a higher viscosity of the continuous phase polymer and a higher shear rate [185]. The blend viscosity ratio affects the connectivity of the phases and the phase inversion region, creating a co-continuous phase morphology. The phase inversion region becomes smaller when the interfacial tension γ increases and is shifted along the concentration axis by varying the viscosity ratio of the blend constituents [186].

In general, in mixing of different rubbers, the goal is to achieve a phase separated blend that combines the properties of the original polymers. The dynamic-mechanical properties of such blends can be considered to be some kind of superposition of the properties of the single phases. Obviously, this superposition is not additive but depends on the morphology and local stiffness and the local strain of the phases. In the case of filled blends, both of these factors are influenced by the partition of the filler between the phases and the dispersion in the discrete phases.

6.7.2 Filler Partition

Filler partition (or distribution) in polymer blends is governed by the degree of affinity of the polymers to a given filler. Due to preferential affinity, chains settle down on the filler surface as long as an adsorption enthalpy is gained. Therefore, filler partition is controlled by all factors that determine adsorption of polymers on filler surfaces. Without leading to a state of equilibrium, the result of this distribution process can express the balance of the filler-polymer interactions. In the case of tire applications in particular, it is of the utmost interest to obtain reliable information on filler distribution, as these can trigger and cause changes in the mechanical-dynamic properties of the materials and their strengths, which are difficult to control [187]. In addition, the mixing procedure and process conditions employed can considerably influence the kinetics of filler partition. Thus, the tendency of filler partition can be predicted from "competing" adsorption of polymers on fillers (see Table 6.1).

The phenomenon has to be considered for two groups of rubber blends. One group consists of blends with high interfacial tension formed by polymers with strongly differing polarity. Besides the morphology with large domains, the difference in polarity leads to a preferential adsorption of one polymer on the filler surface. As a result, the filler can be partitioned almost quantitatively towards the better interacting polymer. The other group consists of blends made from more compatible polymers that exhibit similar δ parameters, small interfacial tension, and demonstrate also similar affinity towards the particular filler. Consequently, the filler partition becomes more balanced. For these blends, determination of a meaningful filler distribution is a rather difficult task.

6.7.3 Evaluation of Filler Distribution

The analytical evaluation of filler distribution makes use of specific contributions of compound ingredients to the physical behavior of phases. That includes the optical contrast, dynamic-mechanical moduli, the composition of bound rubber and others. There are, however, several factors that make the evaluation of filler distribution a rather difficult task. Two factors facilitate the proof and counting of filler distributions in blends: The large interfacial tension between the rubbers that causes large polymer domains and, in general, good phase contrast that can be seen in optical investigations and in mechanical measurements. The other is the interaction potential of chain segments, i.e., expressed by the δ parameter of the polymer, that governs the affinity towards the filler.

Microscopy

The most direct way to gain information about filler distribution is microscopy. The success of optical light microscopy is limited to blend systems containing highly incompatible polymers. Though such blends show quite good optical contrast between the phases and large domains of the phases (µm scale), the limitations of optical microscopy are related to high filler contents and poor resolution of the polymer phases. When the difference in polarity between the polymers and the interfacial tension decrease, the size of the phase separated domains becomes smaller, hence the evaluation of the filler distribution becomes difficult. Transmission electron microscopy (TEM) examination on ultrathin sections delivers qualitatively usable images – albeit at considerable experimental expense – provided the filler exhibits a preferential affinity for one of the two rubber phases and the filler content is low [188]. The quantitative evaluation is also difficult for morphologically uncomplicated boundary cases and – due to the small number of specimens – hardly representative [189]. Above the percolation limit there is practically no chance to evaluate the filler distribution using TEM.

Dynamic-Mechanical Analysis

Filler-dependent changes in the dynamic-mechanical properties of filled blends can be used at a much lower expense and without major restrictions in the temperature range by allowing for a two-phase character of blends. The observation that tan δ decrease with the filler loading was utilized to estimate the distribution of filler between separate phases of immiscible blends, i.e., NR/ENR [190]. However, these results could not be reproduced, mainly because G' and G'' are varying independently as a function of mixing time. The use of the filler-dependent increase of the loss modulus G'' of the individual phases in the glass transition regime in order to determine filler distribution has proven of value in the case of filled blends [191]. The method is based on a mechanical criterion: the linear increase of T'_{max} as a function of the filler volume fraction. Tracing calibration curves $G''_{max}(\phi)$ prior to the evaluation, the filler content in each blend phase represented by a resolved damping maximum is determined by the relative increase of the loss modulus in each phase [192].

For more precise determinations, a curve fitting procedure was developed [193]. By adjusting the amplitude of the relative damping signals and considering the shift in T_g due to polymer compatibility, the experimental curve $G''(T)$ of the blend can be perfectly fitted for temperatures below the first T_g and above the second T_g. The difference between the curve calculated from the addition of the signals for the two filled polymers and the experimental curve appearing in the region between the two damping maxima is attributed to the interface that contains a given amount of filler [194]. The mass balance allows the determination of the filler loading in the discrete rubber phases and in the phase boundary layer. This can be of use if the filler affinities for the two polymers are similar.

Bound Rubber Composition

To reduce uncertainties that arise from the optoelectronic assessment, it was proposed to determine the filler distribution with the help of the composition of bound rubber [195]. It follows that the relative percentages of polymer in the bound rubber should reflect the filler distribution in the blend. Pyrolysis GC was suggested as a suitable method in conjunction with TEM and thermogravimetric analysis (TGA). More recently, the method was applied to evaluate the filler partition in blends of SBR, BR, and NR using representative markers for each polymer. The quantitative determination follows from the ratio of marker peaks in the chromatogram [196]. On the one hand, the difficulty of this method consists in the determination of the bound rubber itself. Often, when a solvent is used to extract the non-adsorbed polymer from the mix, one polymer will be preferentially extracted. Therefore, the composition of bound rubber does not reflect the real polymeric environment of the filler particles. In addition, the lack of adequate markers, the difficult calibration, and the poor precision and reproducibility of GC makes the use of pyrolysis GC impractical and questionable. However, the method is still considered attractive for quantifying CB distribution in blends, particularly for SBR/BR, for which good markers exist.

Crystallinity Loss

It is known that the crystallization enthalpy of semi-crystalline rubbers is reduced by the presence of fillers. The adsorption of the semi-crystalline polymer to the filler surface interferes with the build-up of a crystalline lattice in the carbon black-polymer boundary layer. The number and probably the size of the crystallites formed are consequently reduced in direct proportion to the filler-polymer contact surface. This concept was used to determine filler partition in blend systems containing polymers with very different polarity [197].

6.7.4 Distribution in Blends with Different Polymer Polarity

In blends made from poly-olefines (saturated chain backbone) and poly-dienes (unsaturated chain backbone) a preferential distribution of CB into the unsaturated polymer is always observed. The blend CIIR/NR serves as an example, where the CB particles (dark phase) are mostly located in the NR phase (gray) and to some extent in the vicinity of the phase boundary between the NR and the CIIR phase (bright), while there is practically no CB in the CIIR phase (Fig. 6.24 a) [198].

By determining the crystallinity loss it was confirmed that the distribution of CB in CIIR/BR blends always tends to favor the BR phase and that filler transfer takes place (see the following) [197]. It has been proven for a large number of NR blends that distribution of CBs corresponds qualitatively to the physical-chemical affinity between the filler and the rubber. The main CB content is always in the rubber with the higher degree of unsaturation when blends of IIR, CIIR, or EPDM with polydienes are considered [199].

By way of analogy, in blends made from rubbers containing polar groups and with non-polar rubbers, a preferential distribution of polar fillers towards the polar polymer is observed [200]. Regardless of viscosity and viscosity ratio of the polymers in NBR/EPDM compounds, silica is distributed quantitatively in NBR, (Fig. 6.24 b). By treating the silica with a monofuctional (or bifunctional) silane, the partition can be effectively controlled. Due to the polarophobic effect of the silane moieties, the partition of the filler can be even reversed. In case of an incomplete surface modification, the filler surface achieves an amphiphilic character and is mostly located at the polymer-polymer interface (Fig. 6.24 c), acting as a solid phase compatibilizer and leading to a reduction of the polymer domain size by a factor of 50 [200]. The partition of the filler was quantified for this blend system using dynamic mechanical analysis [191]. It was shown that due to filler transfer similar partition is achieved by prolonged masterbatch blending.

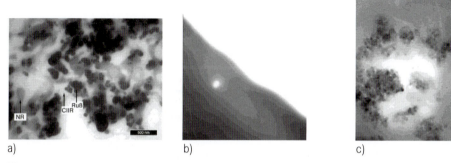

Figure 6.24 TEM-images of a) CIIR/NR-blend containing CB; b) EPDM/NBR-blends containing precipitated silica and c) location of the surface modified silica at the polymer-polymer interface

6.7.5 Filler Distribution in Blends with Similar Polarity

Striking differences can occur in the distribution when the nature of the polymer constituents becomes similar; in particular the content of double bonds or specifically interacting groups [196]. The direct investigation by microscopic methods is more difficult, if not impossible, at higher filler loadings. Because of the better compatibility of the components, the polymer domains are smaller and the phase contrast is reduced so that in the presence of fillers, the blend boundaries become virtually indiscernible. However, on the base of the δ-parameter concept, a filler partition can be predicted from the affinity sequence of the polymers for CB [35, 195]:

SBR > BR > CR > NBR > NR > EPDM > IIR

It was reported that CB distribution depends on the molecular weight (or viscosity) of the polymer and is slightly affected by the surface specific area. The CB partition in SBR/NR or BR/NR blends, as determined by analysis of bound rubber [201] or by dynamic mechanical analysis [191], was repeatedly found to be higher in the SBR and BR phase, respectively. This distribution will not significantly change if the blends are produced in solution instead of in the internal mixer. In addition, it was established that CB distribution preference for the SBR phase is due to the affinity of phenyl groups to the adsorption sites of the filler. A more detailed dynamic mechanical analysis reveals the special case of a preferential filler distribution in the interface (Fig. 6.25) [194].

6.7.6 Filler Transfer

Filler transfer always takes place when the filler finds itself first in a lower-affinity phase and then comes into contact with the high-affinity polymer in the kneader's

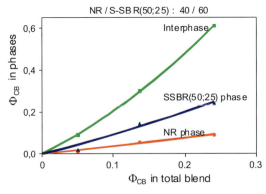

Figure 6.25 Plot of carbon black concentration ϕ_{F_i} in the single phases of S-SBR/NR/CB N234 with blend ratio 4/6 vs. the total carbon black concentration ϕ_F

flow field. The scale of the transfer and its speed is determined by a balance between the interacting forces and the prevailing process parameters. As it was already mentioned, distinct carbon black transfer takes place from the saturated rubber types to the unsaturated and polar rubber types. This direction is hardly influenced by the mixing conditions, which have a much greater influence on the scale of the transfer. A study examining the mixing-time-sensitive transfer of CB to NR/carbon black master batch via SBR indicated a degree of correlation between mixing time and the dispersion of SBR domains in the NR matrix [116]. When mixing times are shortened, the initially CB-free SBR domains are surrounded by agglomerates on the phase boundary. With the increased particle density on the SBR domains, the filler-filler interaction induces a merging of the domains, in the course of which the filler particles move from the phase boundary into the inside of the SBR domains. The carbon black cluster in the domains contributes to the latter's maintaining a hose-like structure.

In the case of the NBR/EPDM system, the extremely varied interactions of the two polymers with silica were shown to affect a quantitative filler transfer within very short mixing times [192]. Despite the EPDM's higher Mooney viscosity, silica is transferred out of this polymer into low-viscous polar NBR in short mixing times so that the filler is quantitatively contained in the NBR phase. The dispersion of silica in the NBR phase corresponds roughly to what is attained in pure nitrile rubber. If in inverted testing the same amount of silica is specified in NBR, no transfer into the EPDM phase occurs. On the other hand, silica rendered hydrophobic, distributes itself evenly in EPDM and NBR. Highly hydrophobic silica is distributed preferentially in the EPDM phase and thus performs like filler with a "paraffinic" surface [155].

In order to locate specific types and amounts of fillers in separate polymer phases, the so called phase mixing can be employed. In a first step, masterbatches with different filler loadings in each blend constituent are prepared. Subsequently, the masterbatches are blended to produce the desired filler partition.

6.7.7 Effects of Filler Distribution

In addition to the physical-chemical and molecular impact of the polymer matrix, rheological effects have a determinant influence on filler distribution. As with all distribution phenomena, interactions between the polymer and the filler will determine the level of distribution at prolonged mixing. The magnitude of change tends to increase in the direction of the larger particle size and structure of the carbon blacks [95]. SBR/BR-blends exhibited the opposite pattern, with viscosity and die swell increasing with higher CB content in the BR. The above mentioned investigations revealed that static modulus (300% elongation) increases slightly at the dis-

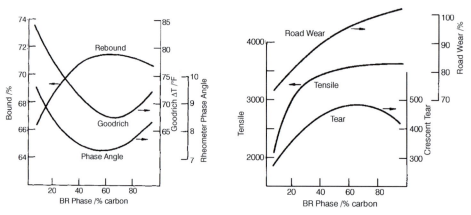

Figure 6.26 Influence of CB distribution in NR/BR blends on a) energy dissipation and hysteresis and b) ultimate properties

tributional extremes for NR/BR. Modulus was highest with 75% of the CB in the BR-phase [95]. Similar results have been reported for dynamic moduli [203]. Low strain modulus increases with an uneven distribution of CB. Concerning the energy dissipation, it was shown for NR/SBR-blends that a higher loading of the NR-phase reduces the heat build-up and increases resilience [204]. It was concluded that hysteresis can be minimized by using CBs with large agglomerate size, broad size distribution, and a selective filler partition.

In blends of NR with either SBR or BR, lower tensile strength was reported with a lower CB loading in the non-strain crystallizing SBR or BR. The most pronounced loss of tensile strength was observed when all of the carbon black was added to the NR-phase [205]. The effect of CB distribution in NR/BR blends on dynamic mechanical and ultimate properties is shown in Fig. 6.26 [97].

The main influences on tear resistance result from (i) small aggregate size and high surface activity, (ii) the higher loading of the continuous phase, and (iii) the polymer with the higher strength compared to the continuous phase [204].

■ References

[1] M. L. Studebaker, *Rubber Chem. Technol.*, **30** (1957), 1400

[2] Kraus, G., "Reinforcement of Elastomers", Wiley-Interscience, New York (1965)

[3] Donnet, J.-B., Bansal, R. C. and Wang, M.-J. (Eds.), "Carbon Black Science and Technology", Marcel Dekker Inc., New York (1993)

[4] H. Palmgren, *Eur. Rubber J.* **156** (1974), 30

[5] J. M. Funt, Mixing of Rubbers, RAPRA, Shrewsburry (1977)

[6] J. L. White, *Int. Polymer Process.* 7(1992), 110

[7] Z. Tadmor and C. C. Gogos, "Principles of Polymer Processing", Wiley, New York (1979)

[8] G. Kühner and M. Voll, in loc. cit. 3, Chp. 1

[9] U. Görl, in "Kautschuktechnologie" (H. Gupta, Ed.) Gupta Verlag (2001), Ch. 9

[10] R. H. Schuster, in "Kautschuktechnologie" (H. Gupta, Ed.) Gupta Verlag (2001), Ch. 8

[11] P. C. Vegvari, W. M. Hess and V. E. Chirico, *Rubber Chem. Technol.* **51** (1978), 817

[12] M. Gerspacher, C. P. O'Farrell, L. Nikiel, H. Y. Yang and F. Le Mehaute, *Rubber Chem. Technol.* **69** (1996), 786

[13] G. Kraus, *Adv. Polym. Sci.*, **8** (1971),156

[14] M. -J. Wang, *Rubber Chem. Technol.*, **71**, (1998), 520

[15] A. N. Gent and H. J. Kim, *Rubber Chem. Technol.*, **51** (1978) 35

[16] S. V. Rao and S. G. Mason, *Nature* **253** (1975), 619

[17] W. M. Hess, *Kautsch. Gummi Kunstst.* **19** (1966), 198

[18] P. V. Dankwerts, *Chem. Eng. Sci.* **2** (1953)

[19] P. M. Lacey, *Chem. Age*, **53** (1945), 119

[20] C. L. Tucker and N. P. Suk, *Polym. Eng. Sci.*, **16** (1976), 657

[21] RELMA Kautsch. Gummi Kunstst.

[22] D. F. Bagster and D. Tomi, *Chem. Eng. Sci.* **29** (1974), 1773

[23] S. Horwatt, D. L. Feke and I. Manas-Zloczower, *Powder Technol.* **72** (1992), 113

[24] B. B. Boonstra and A. I. Medalia, *Rubber Chem. Technol.* **36** (1963), 115

[25] A. I. Medalia, J. Coll, *Interface Sci.* **32** (1970), 115

[26] S. Shiga and M. Furuta, *Rubber Chem. Technol.* **58** (1985), 22

[27] Medalia, A. I., *Rubber Chem. Technol.* **59** (1986), 432

[28] E. G. Harms, *Eur. Rubber J.* **6** (1978), 23

[29] G. Kraus and J. T. Gruver, *J. Polym. Sci.*, **8** (1971), 571

[30] J. H. Hildebrand and R. L. Scott, "The Solubility of Non-Electrolytes", Dover, New York (1964)

[31] R. H. Schuster, H. Gräter and H.-J. Cantow, *Macromolecules* **17** (1984), 619

[32] P. A. Small, *J. Appl. Chem.* **3** (1953)

[33] S. Krause, in "Polmyer Blends" (D. R. Paul and S. Newman, Eds.), Academic Press, Inc., Orlando, Vol. 1, Chp. 2 (1978)

[34] R. H. Schuster, *Angew. Makromol. Chem.* **202/203** (1992),159

[35] R. H. Schuster, H. M. Issel and V. Peterseim, *Rubber Chem. Technol.* **69** (1993), 769

[36] A. I. Medalia, *Rubber Chem. Technol.*, **45** (1972), 1171

[37] P. P. A. Smit and A. K. van der Vegt, *Kautsch. Gummi Kunstst.* **23** (1970)

[38] S. Kaufmann, W. P. Slichter and D. Davis, *J. Polym. Sci., Polym. Phy.* **9** (1971), 829

[39] P. Pissis and D. Fragiadakis, *Macromol. Sci., Part B: Phys.* **46** (2007), 119

[40] D. Bussmann, PhD Thesis (Universität Hannover) 1992

[41] R. H. Schuster, ACS Rubber Division Meeting, Cincinnati (2007)
[42] M. Gerspacher C. P. O'Farrell, L. Nikiel, H. Yang, *Rubber and Plastics News*, Aug. 26 (1996), 39
[43] R. H. Schuster et al., IRC '09, Nürnberg 2009
[44] B. Haidar, ACS Rubber Division Meeting, Las Vegas, (1990)
[45] M. Klüppel, *Adv. Polym. Sci.*, **146** (2003), 1
[46] W. M. Hess and F. P. Ford, *Rubber Chem. Technol.* **36** (1963), 1175
[47] A. I. Medalia and F. A. Heckman, *Carbon* **7** (1969), 567
[48] R. D. Heidenreich, W. M. Hess and L. L. Ban, *J. Appl. Cryst.*, **1** (1968), 1
[49] ASTM D 3849, ASTM Annual Book of Standards, Vol 09.01, p. 630 (1990)
[50] S. Brunauer, P. H. Emmett and E. Teller, *J. Amer. Chem. Soc.*, **40** (1940), 1723
[51] ASTM D 4820, ASTM Annual Book of Standards vol. 9.01, p. 764 (1990)
[52] W. M. Hess, L. L. Ban and G. C. McDonald, *Rubber Chem. Technol.*, **42** (1969), 1209
[53] ASTM D 2414-90, ASTM Annual Book of Standards, Vol. 09.01, p. 434 (1990) 2
[54] M. Gerspacher, C. P. O'Farrell, *Elastomerics*, **123** (1991), 35
[55] C. R. Herd, G. C. McDonald and W. M. Hess, *Rubber Chem. Technol.* **65** (1991), 1
[56] X. Bourrat and A. Oberlin, *Carbon*, **28** (1988),100
[57] F. Ehrburger-Dolle and M. Tence, *Carbon* **28** (1990), 448
[58] B. E. Warren, *J. Chem. Phys.* **2** (1934), 551
[59] L. E. Alexander and E. C. J. Sommer. *J. Chem. Phys.* **23** (1956), 1646
[60] C. W. Schweitzer and G. L. Heller, *Rubber World*, **134** (1956), 855
[61] T. Gruber, W. Zerda and M. Gerspacher, *Carbon* **31** (1993), 1209
[62] R. A. Beebe, J. Briscoe, W. R. Smith and C. B. Wendell, *J. Amer. Chem. Soc.*, **69** (1947)
[63] M.-J. Wang, S. Wolff and J.-B. Donnet, *Rubber Chem. Technol.* **64** (1991), 714
[64] S. Ross and J. P. Olivier "On Physical Adsorption" Wiley & Sons (1964)
[65] A. Schroeder, M. Klüppel, R. H. Schuster, J. Heidberg, *Kautsch. Gummi Kunstst.* **54** (2001), 260
[66] A. Schroeder, M. Klüppel, R. H. Schuster and J. Heidberg, *Carbon*, **40** (2002), 207
[67] S. Yashkin and R. H. Schuster, *Russian Chem. Bulletin*, **52** (2003), 2360
[68] I. L. F. Ray, I. W. Drummond and J. R. Banbury "Developments in Electron Microscopy and Analysis" Acad. Press, London, New York, San Francisco, p. 11 (1976)
[69] S. I. Tang, *Chemtech* **21** (1991), 182
[70] D. Göritz, J. Fröhlich and H. Raab, ACS Rubber Division, Indianapolis (1998)
[71] J.-B. Donnet and E. Custodero, *Carbon* **30** (1992), 813
[72] A. Schroeder, M. Klüppel, R. H. Schuster, *Kautsch. Gummi Kunstst.*, **53** (2000), 257
[73] G. Heinrich und T. A. Vilgis, *Macromolecules* **26** (1993), 1109
[74] J. Ziegler and R. H. Schuster, *Kautsch. Gummi Kunstst.* **56** (2003), 159
[75] R. Ditmar, "Analyse des Kautschuks", Hartben, Leipzig and Vienna (1909)

[76] H. A. Depew and I. R. Ruby, *Ind. Eng. Chem.* **12** (1920), 1156

[77] W. B. Wiegand, *Can. Chem. Metall* **10** (1926), 251

[78] M. Gerspacher, H. Y. Yang and C. P. O'Farrell, ITEC '96, Akron (OH), Sept. (1996)

[79] A. I. Medalia, *Rubber Age* **97** (1965), 82

[80] C. H. Leigh-Dugmore, *Rubber Chem. Technol.* **29** (1956), 1303

[81] Schuster, R. H., Geisler, H. and Boller, F., *Gummi, Asbest Kunstst.*, **96** (1996), 816

[82] W. M. Hess, V. E. Chirico and P. C. Vegvari, *Elastomerics* **112** (1980), 24

[83] H. Geisler, Proc. 3rd Rubber Fall Colloquium DIK Hannover (1998)

[84] Gerspacher, M., Nikiel, L., Yang, H. H. and O'Farrell, C. P., *Rubber Chem. Technol.* **71** (1998), 17

[85] N. Probst in loc. cit. [3] chp. 8, p. 271

[86] N. Probst, and J.-B. Donnet, "Proc. Conf. Conducteurs Granulaires" Paris, Oct. 10, 1990

[87] K. Rajeshwar, L. Nikiel, M. Gerspacher, and C. P. O'Farrell, ACS-Rubber Division Meeting, Orlando, Florida, Sept. 21. - 24., 1999

[88] H. Lorenz, J. Fritsche, A. Das, K. W. Stoekelhuber, R. Jurk, G. Heinrich and M. Klüppel, *Comp. Sci. Technol* **69** (2009), 2143

[89] M. von Ardenne and D. Beisler, *Rubber Chem. Technol.* **14** (1941), 15

[90] A. M. Gessler, *Rubber Age* **94** (1964), 750

[91] M. M. Chapias, M. Polly and R. H. Schulz, *Rubber Chem. Technol.* **28** (1955), 253

[92] R. W. Smith and J. C. Andries, *Rubber Chem. Technol.* **47** (1974), 64

[93] T. Alshuth, R. H. Schuster, and S. Kämmer, *Kautsch. Gummi Kunstst.* **47** (1994), 703

[94] G. Binning and H. Rohrer, *Helv. Phys. Acta*, **55** (1982), 726

[95] W. M. Hess, P. C. Vegvari and R. A. Swor, *Rubber Chem. Technol.* **58** (1985), 350

[96] B. Oppermann, Dissertation, Universität Hannover (1994)

[97] Patent Oppermann

[98] R. H. Schuster, J. Schramm and M. Klüppel, ACS Rubber Division Meeting, Indianapolis (1998)

[99] U. Görl and K. H. Nordsiek, *Kautsch. Gummi Kunstst.* **51** (1998), 200

[100] R. Uphus, O. Skibba, R. H. Schuster and U. Görl, *Kautsch. Gummi Kunstst.* **55** (2000), 279

[101] R. H. Schuster, M. Bogun, L. Muresan, ACS Rubber Division, Pittsburgh, Oct. (2002)

[102] M. A. Marby, F. H. Rumpf, J. Z. Podobnik, S. A. Westveer, A. C. Morgan, B. Chung and M. J. Andrew, US Pat. 6,048,923 (to Cabot Corp. 2000)

[103] M.-J. Wang, T. Wang, Y. L. Wong, J. Shell and K. Mahmud, *Kautsch. Gummi Kunstst.* **55** (2002), 388

[104] M. A. Mabry, F. H. Rumpf, J. Z. Podobnik, S. A. Westveer, A. C. Morgan, B. Chung, and M. J. Andrew, USP 6,048,923 (to Cabot Corporation, 2000).

[105] R. H. Schuster, K. Brandt and L. Schneider, German Pat. 10 2007 048 995.3

[106] R. H. Schuster, Proc. IRC 09, Nürnberg (2009)
[107] C. Wrana "Introduction to Polymer Physics" Lanxess AG, Leverkusen (2009)
[108] M. L. Frisch, R. Simha and F. E. Eirich, *J. Phys. Chem.* **57** (1953), 584
[109] L. Mullins, *Rubber Chem. Technol.*, **42** (1969), 339
[110] R. H. Schuster, D. Bussmann, H. Geisler, 2nd Conference on Carbon Black, Mulhouse 1991
[111] E. M. Dannenberg, *Rubber Chem. Technol.*, **59** (1986), 512
[112] U. Skodzaij, PhD Thesis (Universität Karlsruhe) 2001
[113] C. M. Blow, *Polymer*, **14** (1973), 309
[114] G. R. Cotten, *Rubber Chem. Technol.*, **48** (1975), 548
[115] A. M. Gessler, Proc. Intern. Rubber Conf., Brighton, May 1967
[116] V. Peterseim, PhD Thesis (Universität Hannover] 1995
[117] J. Ferry, "Viscoelastic Properties of Elastomers", John Wiley and Sons, New York (1980)
[118] R. H. Schuster, Proc. IRC 07, Yokohama (2007)
[119] T. L. A. Rocha, R. H. Schuster, M. Jacobi, and D. Samios, *Kautsch. Gummi Kusnst.* **57** (2004), 377
[120] G. Thielen, *Kautsch. Gummi Kunstst.* XX (2008), 377
[121] E. Dannenberg, *RCT* **59** (1986), 512
[122] Hess, W. M., Herd, C. R. and Vegvari, P. C., *Rubber Chem. Technol.*, **66** (1993), 329
[123] W. M. Hess, L. L. Ban, F. J. Eckert and V. E. Chirico, *Rubber Chem. Technol.*, **41** (1968), 356
[124] L. L. Ban, W. M. Hess and L. A. Papazian, *Rubber Chem. Technol.* **47** (1974), 858
[125] M. Gerspacher, *Kautsch. Gummi Kunstst.* XX (2009), 233
[126] J. J. Brennan, T. E. Jermin and B. B. Boonstra, *J. Appl. Polym. Sci.*, **8** (1964), 2687
[127] T. L. A. C. Rocha, C. Rosca, J. Ziegler and R. H. Schuster, *Kautsch. Gummi Kunstst.*, **58** (2005), 22
[128] W. Niedermeier et al., *Kautsch. Gummi Kunstst.* **47** (1994), 799
[129] M. Müller and R. H. Schuster, Proc. IRC '03, Nürnberg (2003)
[130] N. Tricas, S. Borros and R. H. Schuster, *Kautsch. Gummi Kunstst.* **58** (2005), 211
[131] G. A. Joyce and E. L. Little, US Pat. 5,708,055 (to Columbian Chemical Co. 1998)
[132] B. Kershkerian, M. K. Georges and S. V. Drappel, US Pat. 5,545,504 (to Xerox, 1996)
[133] J. A. Belmont, US Pat. 5,554,738 (to Cabot Corp., 1996)
[134] K. Mahmud, M.-J. Wang, and R. A. Francis, USP 5,830,930 (to Cabot Corporation, Nov. 3, 1998)
[135] M.-J. Wang, Y. Kutsovsky, P. Zhang, G. Mehos, L. J. Murphy, and K. Mahmud, *Kautsch. Gummi Kunstst.* **55** (2002), 33
[136] A. Y. Coran and J.-B. Donnet, *Rubber Chem. Technol.* **57** (1984), 959
[137] G. G. A. Böhm and M. N. Ngyen, *J. Appl. Polymer Sci.* **55** (1995), 1041

[138] T. Wang, M.-J. Wang, J. Shell and N. Tokita, *Kautsch. Gummi Kunstst.* **53** (2000), 497

[139] J. Meier and R. H. Schuster, ACS Rubber Division Meeting, Rhode Island (2001)

[140] W. M. Hess, R. A. Swor and E. J. Micek, *Rubber Chem. Technol.*, **57** (1984), 959

[141] E. Guth, R. Simha and O. Gold, *Kolloid Z.*, **74** (1936), 266

[142] E. Guth and O. Gold, *Phys. Rev.* **53** (1938), 322

[143] I. Pliskin and N. Tokita, *J. Appl. Polym. Sci.* **16** (1972), 473

[144] R. W. Sambrock, *J. Inst. Rubber Ind.* **4** (1970), 210

[145] H. M. Smallwood, *J. Appl. Phys.* **16** (1945), 758

[146] A. I. Medalia, *Rubber Chem. Technol.* **45** (1972), 959

[147] G. Kraus, *Rubber Chem. Technol.* **51** (1978), 297

[148] S. Wolff and M.-J. Wang, *Rubber Chem. Technol.* **65** (1992), 329

[149] R. J. Hopper, *Rubber Chem. Technol.* **40** (1967), 463

[150] E. M. Dannenberg and C. A. Stokes, *Rubber Chem. Technol.* **41** (1949), 821

[151] A. I. Medalia, *Rubber Chem. Technol.* **51** (1978), 437

[152] M. Gerspacher, loc. Cit. [3] chp. 11

[153] E. R. Fitzgerald, *Rubber Chem. Technol.*, **55** (1982), 1597

[154] M. Klüppel, R. H. Schuster and J. Schaper, *Gummi Fasern Kunstst.* **51** (1998), 508

[155] J. Ziegler and R. H. Schuster, ACS Rubber Division, Pittsburgh (2003)

[156] Klüppel, M. and Heinrich, G., Rubber Chem. Technol. 68 (1995), 623

[157] Klüppel, M., Schuster, R. H. and Heinrich, G., *Rubber Chem. Technol.* **70** (1996), 243

[158] R. H. Schuster, Material Research Society, Boston, Nov. 2002

[159] A. R. Payne, *J. Appl. Polym. Sci.* **6** (1962), 57

[160] A. R. Payne, *J. Appl. Polym. Sci.* **8** (1964), 2661

[161] G. Kraus, *J. Appl. Polym. Sci.* **75** (1984), 329

[162] D. Göritz, *Kautsch. Gummi Kunstst.* **49** (1996), 18

[163] H. Geisler "Weiterbildungsstudium Kautschuktechnologie", Universität Hannover

[164] M. Jacobi, M. V. Braum, T. L. A. C. Rocha and R. H. Schuster, *J. Appl. Polym. Sci.* **104** (2007), 2377

[165] T. L. A. C. Rocha, C. Rosca, R. H. Schuster, M. M. Jacobi, *J. Appl. Polym. Sci.* **104** (2007), 2377

[166] K. A. Grosch, J. A. C. Harwood and A. R. Payne, *Nature* (London) **212** (1966), 497

[167] L. Mullins, *Trans. Inst. Rubber Ind.* **35** (1959), 3927

[168] A. N. Gent in "Science and Technology of Rubber" (F. R. Eirich, Ed.) Acad. Press, New York (1978) Ch. 10

[169] A. N. Gent, *J. Mater. Sci.*, **15** (1980), 2884

[170] A. Goldberg, D. R. Lesner and J. Patt, *Rubber Chem. Technol.* **62** (1989), 272

[171] H. Geisler and R. H. Schuster, unpublished data

[172] G. Kraus, *Rubber Chem. Technol.* **38** (1965), 284

[173] W. M. Hess, F. Lyon and K. A. Burgess, *Kautsch. Gummi Kunstst.* **20** (1967)

[174] G. J. Lake and P. B. Lindley, *J. Appl. Polymer Sci.* **8** (1964), 707

[175] G. Kraus in Adv. Polym. Sci., (H.-J. Cantow et al. Eds) Springer-Verlag, Berlin, Vol. 8, p. 155 (1975)

[176] E. T. McDonel, K. C. Baranwal and J. Andries in "Polymer Blends", (D. R. Paul and S. Newman, Eds.) Acad. Press, New York, San Francisco, London, Vol. II, Ch. 19

[177] C. M. Roland, *Rubber Chem. Technol.* **62** (1989), 456

[178] O. Olabisi, L. M. Robeson and M. T. Shaw, "Polymer-Polymer Miscibility" Acad. Press, Inc., New York (1979) Ch. 2

[179] R. H. Schuster "Weiterbildungsstudium Kautschuktechnologie", Univ. Hannover

[180] J. E. Kruse, *Rubber Chem. Technol.* **46** (1973), 653

[181] J. Lohmar, Proc. Int. Rubber Conf. Stuttgart, Germany (1985)

[182] K. Kjoller, Digital Instruments, Inc. Santa Barbara, CA

[183] W. Waddell and A. Tsou, *Kautsch. Gummi Kunstst.* **55** (2002) 382

[184] R. H. Schuster, Proc. Int. Rubber Conf., Brighton (1990)

[185] R. H. Schuster, Proc. IUPAC, Cleveland (1992)

[186] G. N. Averopoulos, R. C. Weissert, P. H. Biddison und G. G. A. Böhm, *Rubber Chem. Technol.*, **49** (1976), 93

[187] S. Wolff, 129th Meeting ACS -Rubber Division, New York (1986)]

[188] H. Takino, S. Iuoda, T. Okazaki, T. Sakashita, *Kautsch. Gummi Kunstst.* **43** (1990), 761

[189] R. W. Smith, J. C. Andries, *Rubber Chem. Technol.* **47** (1964), 64

[190] S. Maiti, S. K. De, *Rubber Chem. Technol.* **65** (1992), 293

[191] M. Klüppel, R. H. Schuster, J. Schaper, *Rubber Chem. Technol.* **72** (1999), 91

[192] J. Ziegler, R. H. Schuster, *Kautsch. Gummi Kunstst.* **61** (2008), 510

[193] J. Meier, M. Klüppel, R. H. Schuster, *Kautsch. Gummi Kunstst.* **53** (2000), 663

[194] J. Meier, M. Klüppel, R. H. Schuster, *Kautsch. Gummi Kunstst.* **58** (2005), 82

[195] J. E. Callan, W. M. Hess and C. E. Scott, *Rubber Chem. Technol.* **44** (1971), 814

[196] G. Cotton and L. J. Murphy, *Kautsch. Gummi Kunstst.* **41** (1988), 54

[197] A. K. Sircar and T. G. Lammond, *Rubber Chem. Technol.* **46** (1973), 178

[198] J. Schaube, PhD Thesis, Universität Hannover (2003)

[199] J. E. Callan, B. Topcik and F. P. Ford, *Rubber World*, **151** (1965), 60

[200] F. Boller, H. Geisler, R. H. Schuster, *Kautsch. Gummi Kunstst.* **44** (1991), 116

[201] G. R. Cotton and L. J. Murphy, Proc. Int. Rubber Conf., Harrogate '87 (paper 23A)

[202] M. Klüppel, J. Schaper, R. H. Schuster, ACS Rubber Division, Anaheim (May 1997)

[203] E. A. Meinecke and M. I. Taftaf, *Rubber Chem. Technol.* **61** (1988), 534

[204] W. M. Hess and V. E. Chirico, *Rubber Chem. Technol.* **50** (1977), 301

[205] A. Y. Coran, *Rubber Chem. Technol.* **64** (1991), 801

Index

A

abrasion resistance 20
adsorption 184, 201, 207
affinity 184
agglomerate break-up 180
agglomerates 110, 112
amorphous 76

B

baking behavior 118
base plate 2, 43
batch off 167, 168
batch temperature 126
batch-to-batch fluctuations 143
Black Incorporation Time 54
black scorch 76
bulk density 125
bursting pressure 131
butadiene rubber 84
butyl rubber 86

C

capillary rheometer 144
carbon black 109, 110, 124, 173, 208
carbon black dispersion 149
chain length 204
chain mobility 185
chain stiffness 204
chloroprene rubber 78
cold mastication 72
compatibility 223
compound flow 28
compounding 141
contact pressure 23
contact surface 185
conveying 116
cooling 199
cooling chamber 11
cooling channels 20
cooling pipes 13
cooling surfaces 6
corrosion 20
crack propagation 174
cracks 20
critical agglomerate radius 181
crosslinking 100
crystallization 71, 75, 78
crystallization enthalpy 226

D

DC method 194
dead spots 52, 68
delamination 156
delivery form 127
DIAS 132
digital ram position control 16
disaggregation 126
dispersion 52, 88, 95, 103, 113, 174
dispersion kinetics 209
dispersion measurements 189
dispersive mixing 48, 175
distribution 88, 103, 173
distributive mixing 47, 177
drop door 28, 44, 45, 90
dust 21
dust seals 22, 24
dust settling 11
dynamic crack growth 221
dynamic cut growth resistance 222

E

Electrical resistivity 193
EPDM 75, 108
EPDM long chain branching 121
ESBR 80
extrusion 160

F

feeding hopper 1, 11
feeding temperatures 56
feeding unit 87
fibres 91
filler dispersion 154, 175, 179, 200

filler distribution 222, 229
filler incorporation 151
filler morphology 205
filler partition 224
filler pellets 181
fillers 173
filler transfer 228
fill factor 53, 55, 66, 85, 87, 88, 89, 93
final mixing 67
fines 115
fines content 124
fingerprint 53
first batch effects 59
fluor rubber 86
four-wing rotors 33
free volumes 6

H

hard coating 19, 20
HDSC-rotor 37
homogeneity 177, 178
hydraulic feeding 13
hydraulic rams 61
hydrophobation 97, 98, 99

I

Impedance spectroscopy 194
incorporation 79
injection molding 157
injection valves 29
intake behavior 7, 85
interface 186
interlocking technology 40
intermeshing rotors 5
internal mixer 1
internal surface 6
inverse gas chromatography 188

J

jamming 63

L

laminar mixing 49
latch 44
loss modulus 216
lubricant 83
lubrication film 138
lubrication oil 23

M

macro-dispersion 189, 190, 198, 211
maintenance 26
masterbatch 66, 74
mastication 71, 151
mastication phase 56
MDSC-rotor 38
mechanical scanning microscopy 192
micro-dispersion 193, 198, 206
microscopy 225
milling process 162
mineral fillers 86
miscibility 223
mixer configuration 2
mixing chamber 2, 18, 29
mixing process 151
mixing quality 108
mixing quality index 179, 183
mixing time 198
molecular weight 203
Mooney viscometer 144, 145
Mooney viscosity 84, 155, 157
morphological structure 173
Moving Die Rheometer 131

N

natural rubber 71
NBR 84
noise 14
non-dispersed filler 176
N-rotor 34

O

oil absorption 64, 83
oil addition 156
oil incorporation 153, 212
oil injection 68, 83
operating costs 15
optical roughness measurements 191
organosilanes 95

P

particle shape 186
particle size 186
pellet hardness 111, 113, 125
peptizers 73
percolation 182
PES-technology 41
phase-transition 67
plasticizer oil 129
plasticizers 29, 138, 155
plug 115
pneumatic feeding 12
polarity 227
polarizability 184
polyethylene 72
poor mixing 178
power consumption 124, 126, 197
power curves 61
power demand 56, 92
pre-mastication 67
pre-scorch 102
process conditions 15
processing window 60

Index

process parameters 57, 144

Q

quality assurance 134, 137
quality parameters 121, 135
quasi-stationary equilibrium 59

R

ram 1, 11
ram cleaning 11, 68
ram force 14
ram lifts 68
ram movement 14
ram position 55, 90
ram position control 3, 17, 139
ram pressure 3, 13, 60, 90
ramrod 12
ram seating 55, 66
ram setback position 66
ram speed 16
randomization 177
raw material 108
raw material properties 55
re-agglomeration 211, 215
reflectometry 193
reinforcement 95, 103
reject rates 108, 132
RELMA 128, 132, 133, 145
rheological properties 213
rheovulcameter 145
rolling bank 50
rolling mill 163
rotor bearings 31
rotors 30
rotor speed 52, 58, 66, 91, 198
roughness 192
rubber process analyzer 98, 145, 147

S

SBR 80, 83
silanization 10, 81, 100, 138, 211
silica 95, 97, 173, 189, 208, 211, 219
silica compounds 9
silica-silane reaction 99
siloxane bonds 96
6-wing rotor 38
specific surface area 110, 205
SSBR 80
standard-rotor 34
stearic acid 72
sticking 90
stock temperature 197
storage modulus 217
storage time 72
strain amplitude 217
ST rotor 35
sulphur 76, 127
sulphur silanes 102
surface activity 173, 187, 188, 207
surface area 5
surface defects 76
surface modification 219
surface roughness 188

T

tack 72
tandem mixer 8
tangential internal mixers 32
tangential mixers 4
tangential rotors 3, 4
temperature control 5
temperature probes 27

temperature sensors 26
tensile strength 219
thermal boundaries 59
throughput 52
toggle 45
TOPO 132
torque 55, 65, 102, 197
transmission electron microscopy 194
two-wing rotors 33

V

variable clearance 7
viscosity 53, 72, 73, 74, 88, 89, 109, 203
VOC 101
void volume 112, 187
vulcameter 55, 144, 145, 146

W

wear 91
wear plates 90
wear protection 19
weighing 129
weighing tolerance 131
wings 33

X

X-ray scattering 187

Y

yoke 23

Z

zinc oxide 72
zinc stearate 72
ZZ 2 rotor 36

All about Rubber

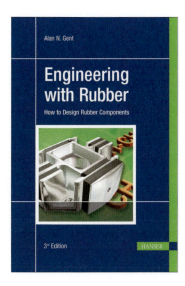

Alan N. Gent
Engineering with Rubber
434 pages.
ISBN 978-3-446-42764-8

This book provides the beginning engineer with the principles of rubber science and technology: what rubber is, how it behaves, and how to design engineering components with rubber.

It introduces the reader to the principles on which successful use of rubber depends and offers solutions to the questions engineers in rubber processing face every day:

- How is an elastomer chosen and a formulation developed
- Why is rubber highly-elastic and relatively strong
- How to estimate the stiffness and the strength of a product
- How to guarantee high quality and durability

More Information on Plastics Books and Magazines:
www.kunststoffe-international.com or www.hanserpublications.com

HANSER

From Structure to Flow Behavior and Back Again.

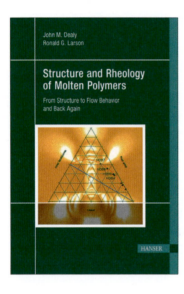

Dealy/Larson
Structure and Rheology of Molten Polymers
530 pages. 130 fig. 12 tab.
ISBN 3-446-21771-1

Several developments have made it possible to predict the detailed molecular structure of a polymer based on its polymerization conditions and to use this knowledge of the structure to predict rheological properties. Soon, it will be possible to use this new knowledge to design a molecular structure having prescribed processability and end-product properties, to specify the catalyst and reaction conditions necessary to produce a polymer having this structure, and to use rheology to verify that the structure desired has, in fact, been produced.

This book provides a detailed summary of state-of-the-art methods for measuring rheological properties and relating them quantitatively to molecular structure.

More Information on Plastics Books and Magazines:
www.kunststoffe-international.com or **www.hanserpublications.com**

The Encyclopedia of Rubber

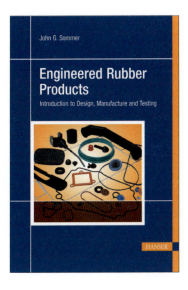

John G. Sommer
Engineered Rubber Products
182 pages.
ISBN 978-3-446-41731-1

The successful manufacture of engineered rubber products is complicated. It involves different disciplines, materials, and types and designs of equipment.

Problems sometimes occur because of less-than-desirable communication among personnel involved in the development and manufacture of rubber products. This book's intent is to improve communication among different disciplines. Using a systems approach, it is further intended to introduce chemists and engineers to the unique capabilities of rubber in a wide range of tire and non-tire products.

More Information on Plastics Books and Magazines:
www.kunststoffe-international.com or **www.hanserpublications.com**

HANSER

Where Fatigue Life is a Concern

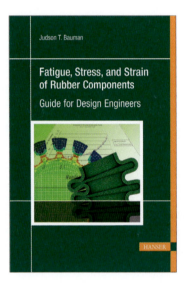

Judson T. Bauman, Ph.D.
Fatigue, Stress, and Strain of Rubber Components
215 pages.
ISBN 978-3-446-41681-9

This text emphasizes the mechanical behavior of elastomers. It discusses the molecular and micro configuration of the rubber matrix and how they produce the observed mechanical behavior.

The fatigue testing of specimens, curve fitting of equations to the test data, and the use of such equations in life prediction are covered compehensively. Stress-strain testing and behavior are covered to the extent relevant to fatigue analysis. Also, the text covers the application of finite element analysis to components to determine high stress points which are vulnerable to fatigue failure.

More Information on Plastics Books and Magazines:
www.kunststoffe-international.com or **www.hanserpublications.com**